MAMMOTH

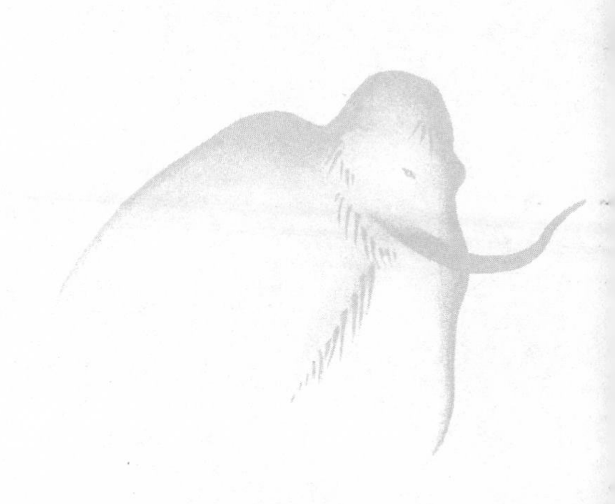

RICHARD STONE

MAMMOTH

THE RESURRECTION OF AN ICE AGE GIANT

PERSEUS PUBLISHING
Cambridge, Massachusetts

Mammoth illustration by Emma Skurnick

Cataloging-in-Publication Data is available from the Library of Congress
isbn 0-7382-0775-6

Perseus Publishing is a member of the Perseus Books Group.
Find us on the World Wide Web at http://www.perseuspublishing.com

Portions of this book first appeared in Discover and Science magazines.
"Song of Khatanga" appeared in "With Love to Khatanga," © Municipal Press-Polygraphic Enterprise "Zelenogorskaya Typographiya," 1999.

Perseus Publishing books are available at special discounts for bulk purchases in the U.S. by corporations, institutions, and other organizations. For more information, please contact the Special Markets Department at the Perseus Books Group, 11 Cambridge Center, Cambridge, MA 02142, or call (617) 252-5298.

Text design by mwdesign
Set in Monotype Bembo in 11.5 pt
First paperback printing, August 2002
1 2 3 4 5 6 7 8 9 10—04 03 02

To Rachel and George

"We knew that all of this was make-believe, and yet it was not."

—CORDWAINER SMITH

Alpha Ralpha Boulevard (1961)

CONTENTS

ACKNOWLEDGMENTS

I could not have written this book without the cooperation of the modern mammoth hunters who allowed me to join them as an observer of their expeditions. My deep gratitude to Kazutoshi Kobayashi of the Creation of Mammoth Association and Bernard Buigues of Mammuthus.

More than three dozen mammoth researchers generously spent time with me. I want to especially thank Larry Agenbroad, Alex Greenwood, Kazufumi Goto, Akira Iritani, Ross MacPhee, Paul Martin, Andrei Sher, and Alexei Tikhonov for their valued contributions to the telling of this story. Most of all, I want to thank Dick and Friedje Mol for their hospitality during my trips to Amsterdam and Sergei and Galina Zimov for the invigorating experience in Cherskii.

Several colleagues deserve credit for indulging me in my mammoth adventures. Colin Norman, Carl Zimmer, and Steve Petranek, in particular, were instrumental in getting me out to Siberia in 1998, 1999, and 2000. Heartfelt thanks to Gabrielle Walker for her wisdom on structuring the book and saving me in a few passages from sounding like the late Jedi Master Yoda.

Tim DeClaire, Andrei Ol'khovatov, Michel Sartori, and Kristen Wainright provided crucial assistance during the researching and writing of the book. I also thank the people at Perseus Publishing for believing in this project from the start.

I could not have written this book without two friends: Amanda Cook, who guided me through the writing process, and Kerry Nugent Wells, who was my muse during every stage of the book's creation.

My plan was to finish *Mammoth* before my son, Aaron, was born on April 1, 2000. I missed that deadline, and might have never finished without the understanding of my wife Mutsumi, who helped research the book and inspired me to keep going when the finish line seemed so far away. Mutsumi's support allowed me to narrowly beat two other special deadlines: The centennial anniversary of the telegram informing Imperial Russia of the discovery of the Berezovka mammoth, and Aaron's first birthday.

Cambridge, UK
26 March 2001

MAMMOTH

RAISING THE DEAD

Cigarette haze filled the kerosene-scented cabin of the vintage Aeroflot helicopter as we flew low across the Taimyr Peninsula in Siberia, but my view from a small, round passenger window was crystal clear. In the late-afternoon sun, treeless hills cast bluish shadows, and ice-packed rivers that snake through a vast frozen tundra glittered like golden blasts of dragon fire. A herd of shaggy reindeer, startled by the *whop-whop-whop* of the strange orange bird overhead, kicked up powdery snow and sprinted by the last larch trees of the Ary Mas reserve, the northernmost forest in the world.

I felt a twinge of anxiety as we approached the barren wilderness site where the veteran Arctic explorer Bernard Buigues and his team had been dropped off two days earlier. I had grown fond of Buigues, watching this modern woolly mammoth hunter in action one week in Siberia in late October 2000. A few days earlier, I had been with Buigues when he flew into a remote village. The indigenous Dolgans who live there—most dressed in western-style winter clothes but a few in traditional reindeer skin jackets and chaps—gathered around and greeted him warmly. Buigues often brings them sacks of flour, shoes for the children, and other supplies to ease their hardscrabble life on the tundra—although on this trip, organized in a rush, he came empty-handed. He asked how they were faring, and the look in his kind blue eyes showed he genuinely cared. Buigues also wanted to know where he could find skeletons of woolly mammoths and—the greatest prize of all—a frozen carcass.

The Dolgans were silent at first. They fear disturbing the remains of a creature revered by the older generations. Yet they divulged to this Westerner, his blond hair graying at the temples, the locations of a few recently spotted bone piles. Buigues had won their trust, but he knew that the Dolgans would never help him unearth a mammoth. Doing so, they believed, would put their lives and their families' lives at risk.

Now I wondered if Buigues had proved the Dolgans prophetic. Cavalier at the end of a long field season, his group had brought only enough food for an overnight stay. Their satellite phone was dead, so they had no way of knowing that the helicopter had been diverted the day before to fix an electricity generator in another village. And we had no idea how Buigues and his team were doing—although weather reports suggested that temperatures on the tundra had plummeted and the winds had grown stronger. Even huddled in their tents, the group would need extra calories to maintain their body temperatures and keep themselves alert and alive. Sitting next to me in the MI-8 helicopter was Christian de Marliave, Buigues's gaunt logistics chief and long-time friend, who appeared unfazed by the delay in rescuing the team. Indeed, it would take far more perilous circumstances to unnerve de Marliave, who had had a delightful time retracing the grueling trek of the early-twentieth-century Antarctic explorer Ernest Shackleton across the mountains of South Georgia Island to reach a whaling station and raise help for his stranded crew.

If Buigues's team was now cold and hungry, it wasn't because he lacked experience in the High Arctic. After the Soviet Union dissolved in 1990, Buigues launched an Arctic tourism business in Khatanga, the biggest town on the Taimyr. His firm in Paris, Cercles Polaires Expéditions, or Cerpolex, caters to daredevil adventurers who want to ski from the tip of the Taimyr Peninsula—Cape Arktichevksy—to the North Pole or to rich tourists who want to sip vodka while standing on the frozen Arctic Ocean. Using Khatanga as a staging ground, Buigues has organized more than twenty expeditions, including a 55-day march on skis from Arktichevsky to the North Pole in 1996. Explaining his fascination with the Arctic, he once told me, "The north is in my soul."

This exotic land captivated me as well. My bird's-eye view revealed a patchwork quilt of frozen ponds and polygonal patterns formed when cycles of severe freezing and thawing fractured the tundra's permanently frozen ground, or permafrost. It's mind-boggling how herds of reindeer can derive enough nutrients from this wasteland.

We approached a large body of water and began to descend. This was Lake Taimyr, its black ice fissured with cracks that had refrozen. I was reminded of the images of Europa, the moon of Jupiter that's thought to be covered by a frozen ocean. Then a brown blob against the black resolved into an ice fisherman's shack, bringing me back to Earth.

The helicopter banked and tents came into view, along with a half-dozen people bundled up against the bitter cold. The copter hovered and settled to the ground, lurching forward a few times until the pilot found a stable landing spot. He cut the engine, and the four of us piled out to joyous greetings. De Marliave, who hides his emotions well, smiled radiantly after hearing that everybody was okay. Maybe he had been worried after all.

The research party, along with the Discovery Channel crew that was filming its exploits, had grown concerned enough to consider sending someone on a 25-mile hike to a shelter equipped with a radio transmitter at Cape Sablera, a lonely outpost with a tattered Soviet hammer and sickle flying outside. And their food had indeed begun to run out —something they could joke about now. "It was terrible," said a cameraman, John Davey. "We ran out of baked beans." After enduring twenty-four hours of hunger and uncertainty, buffeted by winds that dropped the temperature as low as –50 degrees Fahrenheit, everyone was anxious to break camp. Everyone, that is, except the leader.

Buigues smiled at us. "Come, follow me." Without bothering to cover his balding head with the fur-lined hood of his parka, he strode past the tents and down a steep bank to a frozen stream, part of a delta feeding Lake Taimyr. Whipped by gusts, freshly fallen snow curled like wraiths above the flat basin, piling up against the odd rusted-out fuel barrel half buried in the frozen muck. My toes were turning into icicles despite three pairs of socks. Right behind me was de Marliave, pulling a plastic sled loaded with the components of a ground-penetrating radar. We were heading toward what looked like a lone telephone pole set incongruously in the frozen sediments. It was a marker.

As Buigues slowed his pace, the source of his excitement became obvious. Jutting out of the ground were the tops of five chestnut-colored spinal vertebrae, the remains of an emissary from a long-lost world. Buigues got down on his hands and knees and pointed to some tufts of hair. "I didn't expect to find anything when I came here," he said. But in September, before the delta froze, water levels had dropped three feet to expose the top of the mammoth carcass. The previous day, one of the team's scientists had used a steel pick to chip away at the concrete-like delta. A few inches down, he found some frozen flesh from the animal's flank. "We suppose that about half the mammoth is in the permafrost," Buigues said. De Marliave assembled his ground-penetrating radar to try to ascertain just how much more of it they might raise if they were to dig it up.

Buigues is the latest in a long line of mammoth hunters dating back to the days of Peter the Great, the Russian czar who opened Siberia to the world at the turn of the eighteenth century. The French explorer has pioneered a method of airlifting Ice Age animals from their icy tombs in northern Siberia to cold laboratories, where the soil encasing them can be thawed slowly, allowing them to be studied while still frozen. His team hopes that the mammoth remains they unearth will shed light on some of the enduring scientific mysteries surrounding this magnificent beast.

~ ~ ~

Piecing together fossil finds over the last two centuries, scientists have created for our imaginations dioramas filled with wondrous creatures that once walked the earth. No extinct animals are more inspiring than the dinosaurs—but they were never contemporaneous with humans, dying out more than 60 million years before *Homo sapiens* arose. When we imagine enormous beasts that shared the stage with our ancestors, we think of the Ice Age animals that tested human survival. We think of saber-toothed cats and cave lions—but these predators are easily imagined, not so different from modern tigers and lions.

Most entrancing of all the Ice Age creatures are the woolly mammoths. Although no bigger than elephants, their living cousins, mammoths were the biggest animals—and thus the closest to dinosaurs—that our Stone Age ancestors had to square off against. Early on, some scholars thought it was the mammoths that hunted us. As George Turner of the American Philosophical Society wrote in the late eighteenth century, "With a body of unequalled magnitude and strength, it is possible the mammoth may have been at once the terror of the forest and of man!" But it soon became clear that it was the other way around. Our ancestors, armed with spears and guile, hunted these great shaggy beasts. One piece of evidence for this hunt is the remains of igloo-shaped huts built from mammoth bones and tusks on the steppes of southern Ukraine. It's likely that the huts were built thousands of years ago as temporary shelter during mammoth hunting excursions on the steppe.

Mammoths are also special because they are still with us, in body if not in spirit. Because mammoths, along with Pleistocene horses and steppe bison, were the most abundant large animals in Siberia that have since become extinct, their bones are easy to find. But although countless mammoths lay entombed in the Russian tundra, well-preserved specimens are rarer than any gemstone. Of the three dozen or so carcasses ever discovered and retrieved during the brief Siberian summer, most were badly deteriorated or only partial remains. A few exquisite finds, however, have shown us these majestic animals: long narrow heads, downward-sloping hindquarters, small ears, and tusks up to 16 feet long. We know no other prehistoric animal that died out so intimately. And this familiarity forges a connection to a mystical age, a time before written language or the first cities.

With an economy presaging the pencil drawings of Picasso, a few brushstrokes convey the awe in which Stone Age humans held mammoths and their contemporaries. In the Ardèche region of southern France, in a cave closed to the public, two male woolly rhinos battle over a female, locking curving, scythe-like horns nearly as long as their shaggy bodies. Cave lions rush stampeding bison. Two mammoths stand placidly side by side, trunks hanging beneath broad, flat foreheads. These relics are as full of life today as they were 30,000 years ago, when the artists used black manganese dioxide and red ocher, ground and mixed with fat or bone marrow, to sketch the animals on the walls of Grotte Chauvet.

We may know what the mammoth looked like and surmise that our ancestors held it in esteem. Far more compelling are the puzzles that shroud our true relationship with the mammoth. For instance, archaeologists have discovered a cache of mammoth hyoid bones near Krakow, Poland. These bones, which support the tongue, all have cut marks, apparently inflicted by humans who feasted on the mammoth tongues. But why only hyoid bones? Was the tongue a delicacy? After felling and butchering the massive beasts, did Stone Age people use the rest of the mammoth? We know they made tools from the ivory tusks and carved sculptures of mammoths from ivory and slate, possibly to serve as talismans or clan totems. But did they use the hide for foot coverings, the stomach for a water bladder, the downy undercoat for swaddling their babies—as imagined by Jean M. Auel in her novel *Clan of the Cave Bear*?

Although we may never know the answers to these questions, scientists have cracked other longstanding riddles. For instance, we know from studying DNA and tissue samples from frozen mammoth remains that the creatures were cousins, not ancestors, of modern African and Asian elephants. We also know from fossil finds that there were six species of mammoths and that the primeval mammoths originated in Africa before spreading into Europe and Asia.

Eluding scientists, however, is the answer to the most confounding riddle of all: What vanquished the woolly mammoth at the end of the Great Ice Age? This question has captivated Bernard Buigues and his team, who are racing to find evidence frozen in northern Siberia that would support a provocative new hypothesis. If they succeed, they will lay claim to the Holy Grail of mammoth research.

Around 11,000 years ago, the entire mammoth population went into a tailspin, but no one is sure why. Native peoples across Siberia once believed that mammoth bones were the remains of giant rats that dwelled underground and, like vampires, died if exposed to air or sunlight. When Western European scholars heard about the frozen elephant-like creatures in Siberia, they were convinced that these were the carcasses of African elephants swept to the Arctic during the flood described in the Bible's Book of Genesis.

But evidence began to pile up late in the 1700s that mammoths—with thin jawbones, long, twisting tusks, and shaggy hides—were not elephants. So where were they hiding? Many people believed that Noah had brought onto his ark breeding pairs of all Earth's animals; no species had ever just ceased to exist. Even Thomas Jefferson assumed that the mammoth, the bones of which were found throughout the New World, existed in the wilds west of the Mississippi River. This fallacy was shattered during Jefferson's presidency, when the woolly mammoth became the first species ever shown to have gone extinct.

When, in the 1840s, a picture emerged of glaciers advancing deep into Europe and North America during the Great Ice Age, biologists reasoned that the woolly mammoth was a cold-adapted mammal that could not cope with the abrupt transition to a warmer climate when the glaciers receded. A more sophisticated version of this idea holds that during much of the Pleistocene Epoch—about 1.8 million years in which the polar ice sheets advanced and retreated twenty-three times—huge swaths of the Northern Hemisphere were covered by cold, arid grassland: the mammoth steppe. When the world warmed around 11,500 years ago, near the end of the Pleistocene, shifting weather patterns brought more rain and snow. Winter would have become particularly taxing for grazing animals, when mosses and sedges—less nutritious plants that had supplanted the steppe's grasses and herbs—lay buried under snowdrifts. In this scenario, mammoths couldn't get enough to eat, and isolated populations guttered and winked out one by one.

Thirty years ago, a rival hypothesis emerged to pin the blame for the mammoth's disappearance in North America on prehistoric humans who slaughtered the animals for food. Attributing the demise of mammoths across Eurasia to hunters was much more of a stretch, but the idea got European researchers probing whether hunting abetted climate change in dooming the animals.

A spine-chilling new idea appeared in 1997 when Ross MacPhee, the curator of mammals at the American Museum of Natural History in New York, argued that prehistoric hunters, or perhaps their dogs, carried a deadly microbe that tore through mammoth populations with the virulence of the flu and the lethal quality of the Ebola virus. "There was no way the mammoths could fight back," MacPhee says. "This was the killer plague of all time." MacPhee thinks that this efficient killer, or at least its DNA, may still exist in the frozen bones or flesh of the most recently deceased mammoths, so in August 2000, he joined Bernard Buigues on the mammoth hunt on Taimyr to try to prize the "hyperdisease" pathogen, if it exists, from the ice.

~ ~ ~

Tracking down a prehistoric terror is not the only goal of modern mammoth hunters. Finding the frozen mammoths today holds out a fantastic new possibility: What if we could bring one of these fabled creatures back to life? The Japanese scientist Kazufumi Goto believes we can. In a pioneering experiment in 1990, he showed that a bull's dead sperm could be used to fertilize an egg. After implanting the egg in a cow, a calf was born.

If the intact DNA, the genetic code, contained in a sperm head were all that was necessary for fertilization, Goto thought, dead sperm might be used to save endangered species—or to resurrect Ice Age beasts. Goto briefly considered the idea of reviving Ice Age animals like deer or bison, whose modern descendants could serve as surrogate mothers for any breeding experiment using dead sperm. But bringing back a woolly mammoth would be more romantic. Mammoths are so unlike any animal alive today, yet they are closely enough related to elephants that a breeding experiment might just work. In the summer of 1997, Goto assembled a team and struck off for Siberia in search of frozen mammoths.

Not long ago, that sort of wild-eyed adventure and its ultimate goal—bringing the mammoth back to life—would have been dismissed as science fiction, a kind of warm and fuzzy *Jurassic Park*. But stunning advances in reproductive biology are giving scientists a shot at succeeding. They range from the microscopic tools that allow scientists to slip a sperm head into an egg without rupturing its fragile membrane to the technological tour de force of removing an egg's genes and inserting replacement genes from an adult cell—the necessary prelude to making a clone.

If he could find well-preserved tissue, Goto reasoned, he would face a choice: attempt to create a hybrid, with half its DNA from a mammoth and half from an elephant, or create a clone that would be 100 percent mammoth. Producing a hybrid would require injecting mammoth sperm into an Asian elephant egg. (An African elephant, also closely related to the mammoth, could be tried too.) If an embryo were to form from the union, it would be implanted in an Asian elephant cow. It's likely that untold numbers of hybrid embryos would never develop into a fetus. But if one were to flourish and be carried to term, it's conceivable that after around 600 days—the length of time a fetal elephant stays in its mother's womb—a living mammoth would be born.

Cloning a mammoth would be more spectacular—and possible, in theory—with mammoth tissue in which the DNA, a long, spiraling, ladderlike molecule, had not shattered during the freezing process. Scientists could inject the nucleus from a mammoth cell directly into an Asian elephant egg stripped of its own DNA and zap it with electricity to spark fertilization. Such an embryo would also be implanted in an Asian elephant cow.

Scientists are on the verge of bringing extinct species back to life. "I know it sounds unbelievable," says Goto. "But no science can deny our idea."

— — —

This book is about the daring individuals who are penetrating the farthest reaches of Siberia in search of the mammoth. Breathtaking finds from this great mammoth graveyard have ushered in each of the last three centuries. Two hundred years ago, the Adams mammoth provided a glimpse of what the great beast looked like. One hundred years ago, the majestic Berezovka mammoth added more vital clues to its appearance, giving us our present picture of the mammoth.

Today, explorers are mounting the most intense search ever for frozen mammoths. For good reason. While dinosaur hunters must settle for dusty bones, the DNA in these relics long gone, mammoth hunters know they can find the real thing: gnarled plaits of mammoth hair, still attached to the skin, smelling mangy—and alive. Beneath the skin lie the muscles and ligaments and fat, the bones and even organs, all still containing their DNA, which holds the secrets to why a mammoth looks different from an elephant. Today's explorers are turning up new specimens that could at last explain why the woolly mammoth became extinct. They also offer the magical possibility of bringing the king of the Ice Age back to life.

Would our lives somehow be enriched after resurrecting the king? Or are we courting disaster, as the American paleontologist William Berryman Scott, believing the mammoth to have been a ferocious carnivore, suggested in 1887: "The world is a much pleasanter place without them, and we can heartily [say] 'thank heaven that the whole generation is extinct.'" To many people, conjuring up a breathing mammoth is a chilling thought. The Dolgans and other native Siberians who live with the bones of the beast beneath them fear the consequences of bringing it back to life. Even some of the scientists and explorers contend that resurrecting the mammoth is pointless or amoral, considering how alien it might feel in a world much changed from the one of its brethren. End this Promethean quest, they argue, and let the mammoth rest in peace.

That will not be possible. Explorers and scientists are racing to raise the dead. Their goal is to exhume the most intact mammoth ever beheld since our ancestors stalked the hairy beasts. For some, the finish line spells doom; for others, the start of a spectacular scientific drama.

TREASURE OF THE WOODEN HILLS

Looking like a lumberjack in his red plaid cotton shirt and gray skullcap, Anatoly Logachev shifted his bulldozer into gear and plowed into the thawing ground on a warm and sunny morning. It was June 23, 1977, early in the summer strip-mining season at the Susuman gold mine in northeastern Siberia's Kirgilyakh River valley. The night before, the miners had redirected a small stream to thaw and wash away an ice lens covering a promising patch of ground. While working this layer, Logachev's blade caught on what appeared to be a thick, dark root. Climbing out and examining what he'd gashed, Logachev discovered that the root, in fact, was part of a body. He called over the other miners, who brought a hose and starting blasting the concrete-like permafrost with warm water. After a few minutes, enough silt had washed away for them to get a good look at the carcass. Logachev's eyes flashed. "*Mamontyonok!*" he is said to have cried. *A baby mammoth!*

Logachev did not realize, but he was looking at the best-preserved frozen mammoth ever found.

Most miners would probably have discarded the carcass: any distraction that curtailed their operations during the brief Siberian summer would hurt them financially. But Logachev had a hunch that this baby mammoth was special, so he persuaded his fellow miners not to hush up the find. That day the head of the mine cabled the director of the geology institute in the regional capital of Magadan, 180 miles to the east. As word got out, a stream of local officials and residents came to gawk and take photos. Over the next two days the weather grew hotter, and the thawing *mamontyonok* began to stink and draw flies. The miners covered it with ice and put up a makeshift tent.

Finally, three days after the carcass was exposed, three geologists arrived from Magadan. They immediately recognized the specimen's value and moved it to a shelter to protect it better from the sun; later they moved the thawing corpse by car to Magadan, where it was stowed in a freezer. Word filtered up the chain of command to Communist party authorities, who on July 3 dispatched teams from the Institute of Paleontology in Moscow and from the country's leading mammoth house, the Institute of Zoology in Leningrad (so called in the Soviet era, the city's name reverted in 1991 to St. Petersburg). One of the scientists who flew over the eight time zones to Magadan was Nikolai K. Vereshchagin, the zoology institute's main mammoth expert. On his arrival, he marveled at the baby, dubbed Dima after a rivulet several feet from the site. "I laid my hand on its skin and felt a chill," Vereshchagin recalls. "I had touched the Stone Age."

During the examination of the site at the Susuman mine, Vereshchagin congratulated Logachev and presented him with all he was permitted to give: a special watch from the Soviet

Academy of Sciences in reward for reporting the find. (A few years later, the Soviet government insured the specimen for 10 million rubles before sending it to London for an exhibition.) Feeling sorry, Vereshchagin asked the Communist party secretary in Magadan to supplement the measly reward by giving Logachev and his crew another bulldozer. "I was trying to do something for the workers," Vereshchagin says. The functionary refused. But Logachev always remained dear to the scientists' hearts: he later found a place of honor at the Zoological Museum in St. Petersburg, which exhibits a framed sepia-tinted photograph of the miner squinting in the sun in front of his bulldozer. The picture hangs above two glass jars, one containing Dima's heart, the other the *mamontyonok*'s pickled foot-long penis.

Among mammoth researchers, Vereshchagin arouses a range of emotions, from adoration and awe to fear and ridicule—but never apathy. "He's a monster, but I love him," says Alexei Tikhonov, a mammoth hunter at the zoology institute who has spent most of his career working under Vereshchagin and is now part of Bernard Buigues's expedition. Most researchers at the institute give the curmudgeon a wide berth. "He's like a vampire; he comes to the institute and attacks two or three people, usually without reason, before he's satiated," says Tikhonov. "He tries to provoke me, but I don't bite. I have a thick skin." The institute tried to force Vereshchagin into mandatory retirement when he was seventy-five years old, Tikhonov says, but he wouldn't go until he was eighty-five. The dean of mammoth hunters still has an office at the institute, which he shares grudgingly with his junior colleague. "I don't have permission to touch his fossils," Tikhonov says.

I met Vereshchagin on a typical February day in St. Petersburg. Huge wet snowflakes fell from a gray sky. Several Russian navy cadets, dressed all in black, skied through a park of birch trees across the street from St. Isaac's Cathedral, its magnificent golden dome speckled with new snow. After trudging along the slushy sidewalk for a few more minutes I came to the River Neva, the soul of the city. There stood the sea-green Winter Palace, its pilasters marching the length of two football fields. My destination was a pale blue building on the other side of the Neva. While not nearly as ornate as the czar's former home, the Institute of Zoology of the Russian Academy of Sciences also commands a grand view of the river.

Vereshchagin, who had recently celebrated his ninety-first birthday, had made a rare trip in from his dacha near Lake Ladoga, north of the city, to talk with me. Much warmer and more personable than I expected, Vereshchagin showed off trophy fossils not on display in the institute's museum, including the skulls of a cave lion and several steppe bison, two species that were much larger than their modern relatives. With 15 million specimens collected since the days of Peter the Great, the zoology museum, one of the greatest in the world, is forced for space reasons to keep some of its prize artifacts in the back rooms for lack of space.

In Vereshchagin's cluttered office, a pair of ridged mammoth teeth the size of lunchboxes perched precariously on the mountain of papers covering his desk. Portions of three mammoth tusks, each nearly 3 feet long, sat on a bookshelf, their tips curling over the edge like the talons of a giant bird. His gray hair neatly parted on the side, Vereshchagin wore a clean white lab coat, even though he wouldn't be doing any work that day. Tucked in the coat was a navy-blue tie speckled with tiny gold mammoths. Notwithstanding the blood-soaked gauze around his left wrist—his dog had bitten him the day before—the blue-eyed nonagenarian looked remarkably hale. He had little patience for questions, intent instead on telling stories that he deemed important or entertaining. His thoughts eventually wandered back to Dima.

As the scientists in Magadan launched an investigation of Dima, Vereshchagin recalls, photos went out on the newswires of the greatest mammoth find of all time. People around the world were captivated by the sight of a mammoth, not much bigger than a sheep dog, lying on its side on an examination table. The animal's trunk, with the peculiar handlike grip at the end, curled back toward its body.

To their dismay, the Soviet scientists found that Dima had dried out in the days it lay outside thawing and that bacteria had invaded the carcass. Nevertheless, a tussle ensued over the control of Dima that summer, with the Siberian branch of the Soviet Academy of Sciences, based in Novosibirsk, laying claim to the find. But Vereshchagin and the director of the Institute of Geology in Magadan cut a deal to publish a joint scientific report on Dima, and early in August the carcass was flown to Leningrad. In retrospect, Vereshchagin says, "that was a big mistake." The institute's taxidermists bungled the job, inexplicably soaking Dima in benzene before embalming him in paraffin. As a result, the specimen lost nearly all its remaining hair (much of it had shed when it was first thawed at the mine) and the skin turned from light brown to tar black.

Nevertheless, Dima still had a story to tell. Initial estimates based on the geology of the site suggested that the baby mammoth had died around 10,000 years ago, but radiocarbon dating of the specimen itself showed that Dima had taken his final breaths around 30,000 years earlier. Despite its great age and its thawing and refreezing, portions of the 200-pound carcass seemed nearly as fresh as the day the animal died. Vereshchagin's team found a massive blood clot in the mammoth's gut and traces of his mother's milk in its stomach. This and other evidence suggested that Dima was on his own for less than a week before he died. He was probably six to eight months old at the time, which would have put his death in late summer or early autumn.

Why would a baby woolly mammoth be wandering alone on the steppe? A mother would never abandon a healthy calf. Nor was she likely to lose track of him. As mammoths almost surely behaved like elephants, she or a relative would have nudged the straying calf back to the herd. It is also doubtful that the mother had been ambushed. The only predators that dared attack healthy, full-grown mammoths were humans, and it would be another 20,000 years at least before bands of hunters penetrated this far north into Siberia. A cave lion might pounce on a baby or an ailing adult unable to keep up with a herd, but a protective mother would have been out of its league.

More likely, Dima's mother was the victim of a tragic accident. Perhaps she had plunged into a bowl-shaped steppe lake and found herself unable to crawl out, leaving Dima on his own. Covered in straw-colored downy fur, fringed by locks of thicker, reddish and brown hair, the 3-foot-tall calf would have clung to the lakeshore, sucking up dirt and grass with his trunk, which was just beginning to sprout the thick brown hairs and coarser guard hairs of an adult. But he would have craved the nutrients in his mother's milk and grown steadily more famished. His baby fat almost spent after nearly a week on its own, the *mamontyonok* might have then stumbled into a muddy pit. After a struggle to extricate himself, he might have crumpled to his knees, sinking below the surface and suffocating.

Within a few days a cold snap would have frozen the mud and the carcass, and autumn's first snow would have hidden it from scavengers. By the time the ground began to thaw the next summer, silt carried by the spring floods would have filled the depression, entombing the baby's remains for the next 400 centuries.

— — —

Dima wasn't the first mammoth to bring to life a lost world. Three earlier sensational finds by mammoth hunters had taught us crucial lessons. There was a time, not so long ago, when what little information we did have was largely drawn from myth.

The first recorded reference to the mammoth comes from the Chinese, who traded for ivory with the indigenous peoples living to their north, in today's eastern Siberia. According to a classic text called the *Shên I King*, attributed to Tung-fang So in the second century B.C., "In the north ... the *K'i shu* is found beneath the ice, in the midst of the ground. In shape it is like a rat. It eats grass and trees. Its flesh weighs a thousand catties [roughly a thousand pounds], and may be used as dried meat for food.... Wherever its hair may be found, rats are sure to flock together." (*Shu* means rat, or rodent; the origins of *K'i* are obscure.) Tufts of hair were scarce, yet bones were everywhere—but no one realized that they were fossils. The *Shên I King* also notes that eating cold mammoth flesh is good for reducing fevers.

In the Middle Ages, a leading theory held that fossils had nothing to do with life. Instead, they were inorganic *lusus naturae*, freaks of nature, that were forged inside the churning earth, perhaps, or calcified impressions left in the ground by Roman artifacts. This notion persisted into the early 1700s.

Another school of thought held that mammoth bones (and rarer dinosaur bones) were from monsters or mythical giants. One candidate was the sea monster Cetus, which Perseus slew before it could devour Andromeda, his brideto-be, strapped to a rock on the shore as a sacrifice. Other legendary giants came to mind as well, including the hunter Orion and the 19-foot-tall Cimbrian king Teutobochus, whose skeleton, unearthed in France, was later shown to be a mammoth's.

A strange but logical explanation arose for mammoth skulls, which have large cavities where the trunks were once attached. This cavity looks like a single eye socket and may have been the source of the myth of the Cyclops, one-eyed giants that forged thunderbolts for Zeus. Mammoth bones were attributed on occasion to unicorns; they were even venerated as the remains of saints: the church of Valencia in the 1500s would parade through the streets a tooth of Saint Christopher, which was in fact a mammoth molar. Mammoth femurs, meanwhile, masqueraded as the arms of saints.

Most experts attribute the first mention of the mammoth in western Europe to Nicolaas Witsen, a mayor of Amsterdam who visited Moscow in 1665 and learned about the *mammout* (which may have been derived from the Tartar word *mamont*). In his *Noord en Oost Tartarye* (1695), Witsen wrote that the creature, if spotted, "betokens much calamity." Shown the remains of the peculiar animal, the Dutchman described them as dark brown and smelly. The elephant-like teeth, he learned, were found along the banks of Siberian rivers. He speculated that the very large but rarely seen *mammout* was a mystical creature perhaps connected to the behemot, or *behemoth,* the gigantic beast with twisting horns in the Old Testament's Book of Job.

A few years before Witsen's book appeared, the enlightened young czar of Russia, Peter the Great, sent an envoy to the emperor of China who would add flesh to the bones of this mythical ice rat. In the diary of his journey across Siberia, Eberhard Ysbrant Ides described the cultures and lifestyles of the indigenous peoples he met along the way. It didn't take him long to encounter people who knew the mammoth well. According to the Tungus, Yakuts, and Ostyaks, he wrote, the bones and tusks belonged to animals that "continually, or at least by reason of the very hard frosts, mostly live under ground, where they go backwards and forwards." One whiff of fresh air was enough to strike the creature dead—explaining why those unlucky beasts that emerged from their subterranean lairs were never seen alive. The Samoyed people, meanwhile, thought that the mammoth—the "Earth Master"—dwelled in the mountains and would sneak into villages at night to graze on human remains in the shallow cemeteries.

Around that time, mammoth bones were beginning to lose their supernatural mystique in western Europe. When giant bones were unearthed in a German quarry in 1696, investigators with the Collegium Medicum hewed to conventional wisdom and decreed that the remains had belonged to a garden-variety unicorn. Seeking a second opinion, the local count consulted another scholar, Wilhelm Ernst Tentzel. His careful report came to a radical conclusion: the bones were from an ancient elephant.

That idea soon became popular throughout Europe. Intellectuals, including Peter the Great, suggested that the bones in Europe and Siberia belonged to elephants that had strayed from the herds of Alexander the Great or other military commanders said to have ridden elephants into battle. It sounded plausible, but there were far more bones in Siberia than could be explained by a handful of stray elephants. Particularly perplexing was how, if the ivory was so valuable, the tusks of these elephants were buried with the bones and left to rot. Others searched for answers in the Bible, attributing the bones to Job's behemoth. Lending support to the idea, it would seem, was the concept of an underground rat that died when exposed to sunlight; to vanquish the behemoth, it says: "He taketh it with his eyes" (Job 40:23).

It's a nice metaphor, but many scholars balked at the notion of a Siberian monster. Instead, wrote the nineteenth-century paleontologist William Berryman Scott, "that other wonderful solution of all difficulties, Noah's deluge, was called in to account for the anomaly."

The idea was that the elephant carcasses had been washed north during the biblical flood and, after the waters receded, had lodged in fissures in the permafrost. "After this Noachian deluge, the air which was before warm was changed to cold," freezing the bones and preserving them from putrefaction, Ides wrote. However, indigenous Siberians knew little, if anything, about elephants. So when Cossack traders and Russian Orthodox missionaries established settlements in Siberia, a melting pot of beliefs emerged, including a new version of the Noah story with an ignominious end for the mammoths. Two of the ungainly creatures trundled aboard, rocking the ark so violently that Noah, in a panic, beat back the beasts, jettisoned the gangplank, and shoved off without them. A variation was more forgiving toward Noah: the mammoths boarded the ark and survived the deluge, but the moment the heavy beasts disembarked, they sank into the waterlogged ground and disappeared. Such biblical revisionism never made it back to Europe, where people continued to believe that the bones dug up across Europe and in Siberia belonged to elephants.

At the close of the eighteenth century, a great comparative anatomist finally dispelled this notion. Baron Georges Cuvier compared thousands of fossils from diverse species with the bones of modern relatives. In 1796 the Frenchman concluded that the fossils represented no fewer than twenty-three extinct species. Among his revelations, Cuvier argued that the mammoth and woolly rhino were not only different creatures from their African counterparts but also were adapted to life in the Arctic: these were surely not Alexander the Great's elephants. "The great lesson which Cuvier taught the world," wrote Scott about the founder of paleontology, was "that many races of animals were entirely extinct, and that nature's chain of existence had not one, but many missing links."

~ ~ ~

But Cuvier could not put flesh on the mammoth. That task lay to the Scottish botanist Michael Adams, who in 1806 hauled from Siberia to St. Petersburg the first nearly complete skeleton, hide, and fleshy parts of a mammoth. If he had arrived at the site of the find seven years earlier, he might have seen the whole beast frozen and intact.

In late August 1799, a hunter named Ossip Shumakhov took a break from fishing to search for mammoth tusks along the coast of Cape Bykov, where the Lena River forms a delta and drains into the Laptev Sea off north-central Siberia. Shumakhov knew that if he found a freshly thawed mammoth tusk with its ivory in good condition, he might be able to barter it in town to supplement his meager income.

The Arctic summer provided a brief period for Shumakhov and other Tungus people to look for tusks emerging from the Pleistocene sediments; temperatures in September fell below freezing and generally stayed there, along with a blanket of snow, until spring. Tusk scavenging was—and still is—a treacherous occupation. As the coastal tundra thaws, sediment crumbles into the ice-choked sea, where it mixes with the bones of Ice Age animals, wind-blown dust known as loess, and driftwood that the Cossack colonists called *noevchina,* after Noah. (Remains of ancient forests buried in the tundra—called *adamovchina,* since they were thought to date to the time of Adam, in the Bible—are so common that one explorer poetically referred to a series of Arctic bluffs as "the wooden hills.") In the Laptev Sea, the ice pack jams the *noevchina* and other debris against the shore. The slurries, like quicksand, along northern Siberian waterways in summer have been known to trap hapless animals and people.

Along the Laptev's tangled, mucky shoreline, one block of frozen ground struck Shumakhov as odd. He saw a shape protruding from the bank but could not make it out. The next summer Shumakhov noticed two projections but still could not identify them. By 1801, however, one side of the mass had been exposed, and Shumakhov could tell that it was an animal with tusks. When the Tungus chieftain told others in his village about the thawing beast, he got a chilling warning. Years earlier, older members of the community said, a similar monster was liberated from ice on the delta. The man who discovered it—and his entire family—died soon after.

Petrified that he may have brought a curse upon his household, Shumakhov fell seriously ill. But he recovered and, perhaps figuring that he had survived a test, began lusting after the tusks and the goods they might bring. Shumakhov asked his people to waylay any strangers who happened to approach the mammoth's location. The following summer was unusually cool, and the mammoth stayed frozen. After a warm spell in 1803, however, the carcass melted free and toppled onto a sandbank. That winter, Shumakhov sawed off the tusks and bartered them for 50 rubles worth of goods in the closest town, Kuma-Surka.

Although Siberia's vast Ice Age graveyard contains countless bones and tusks, whole carcasses are rare and often badly damaged. Thus, at the turn of the nineteenth century, people had only a vague notion of what the tusks' owners might have looked like. The tusks were later sold to a merchant in

Yakutsk. Based on secondhand descriptions, the merchant drew a fantastic beast with pointy ears, tiny eyes, horse's hooves, bristles along its back, and two curving tusks pointing downward and away from each other, like the dovetail of an anchor. Quite an uncomplimentary rendition of a mammoth, it looked more like a deformed boar.

Michael Adams, who was affiliated with the Imperial Academy of Sciences in St. Petersburg, heard about Shumakhov's beast while traveling to China in 1806 with Count Golovin, Russia's ambassador to Peking. Stopping for a rest in Yakutsk, Adams encountered the merchant, who showed him the unusual drawing. Adams didn't know what to make of the beast, but he was intrigued to learn that the animal, including its flesh, skin, and hair, was still fairly intact. He realized that it would be a scientific coup if he were able to find a whole mammoth at the mouth of the Lena River. Adams sent the drawing and the description of the carcass to his colleagues in St. Petersburg, begged the count to remain in Yakutsk for a while, and set off for Kuma-Surka.

When Adams reached the shore of the Laptev Sea the next month, he found his trophy badly mutilated. The villagers had hacked off the flesh and fed it to their dogs. Bears, wolves, and foxes had nibbled at the leftovers, leaving clean bones. But the carnivores had rejected a valuable consolation prize: most of the hide. Adams ordered several local men to drag the dark gray skin, covered with reddish wool and black hairs and a mane running along the neck, away from the water. Inside the skull he discovered the animal's dried-up brain. Adams also scooped up 36 pounds of loose hair from the depression in the sand left by the carcass. "I found myself in possession of a treasure which amply compensated me for the fatigues and dangers of the journey," he wrote. He forced the local men to haul the skin, skeleton, and hair all the way to Yakutsk by horse-drawn sledge. The tusks had already been cut in pieces and sold, so Adams purchased another pair and, after boiling the bones to remove the ligaments and other crud still covering them, sent the specimen by sledge and train to St. Petersburg.

This find put the wool on the mammoth, suggesting that these creatures were adapted to survive in a harsh Arctic environment. The Adams mammoth, as it came to be called, was also the first mammoth skeleton ever exhibited—incorrectly, it turned out. While overseeing the skeleton's mounting, Adams stuck the tusks in the sockets pointing away from each other—the prevailing view being that the mammoth used its tusks as a pair of scythes for defense and for clearing snow. Without having seen the tusks in their original position, Adams could not have known otherwise.

~ ~ ~

As Europeans grew fond of this big hairy beast and its name came to describe anything gigantic, North Americans began to develop a similar fascination. Mammoths were supposedly spotted dozens of times throughout the 1800s, especially in Alaska. A spate of sightings near the end of the century, however, described the mammoth so vividly and accurately that scientists were inclined to believe it hadn't died out. But this trail proved false as well. The rumors were traced to an innocent exchange between Charles H. Townsend, a naturalist with the United States Fish Commission, and a few of Alaska's Inuit people whose names were never recorded.

When the cutter *Corwin* stopped briefly at Cape Prince of Wales in Alaska's Kotzebue Sound, north of the Bering Strait, the Inuit came aboard to show Townsend some mammoth bones and tusks. They asked him to describe what sort of creature the bones had belonged to—they had never seen one alive. Townsend showed them a woodcut of the Adams mammoth skeleton displayed in St. Petersburg, and he sketched what the mammoth may have looked like in its woolly hide based on the hair and skin recovered by Adams. The copies were taken ashore, where apparently they were recopied many times and distributed across the territory by the peripatetic Inuit. Learning later that some Inuit could draw surprisingly accurate renditions of a mammoth, a newspaper reporter concluded that the Native Americans were producing these sketches from memory, having seen a living mammoth in the Alaskan interior.

In an article in the February 1900 issue of *McClure's* magazine, the paleontologist Frederic A. Lucas argued that it was "utterly improbable" that any living mammoth had been seen on the Alaskan tundra for centuries, and he concluded that there were no live mammoths to be had at any price, anywhere. He ended with this challenge:

> Should any man of means wish to secure enduring fame by showing the world the mammoth as it stood in life a hundred centuries ago, before the dawn of even tradition, he could probably accomplish the result by an expenditure of a far less sum than it would cost to participate in an international yacht race. Who will be the first to dispatch an expedition to seek a frozen mammoth?

The answer came a few months later, when scientists embarked on a mission to retrieve a most spectacular mammoth.

— — —

At the turn of the twentieth century, St. Petersburg was a cosmopolitan city in its golden era. Then in his prime, Peter Carl Fabergé and his artisans were producing their famous jeweled eggs and other lavish works for the court of Nicholas II and for wealthy patrons across Europe. The prima ballerina Anna Pavlova enthralled audiences at the Mariinskiy Theater. And while Pyotr Tchaikovsky had just died—from cholera or, some scholars speculate, from drinking poison after an alleged love affair with his nephew was exposed—his symphonies and operas, such as *Sleeping Beauty,* had become classics at the renowned conservatory. It would be another decade before Grigory Rasputin charmed his way into bourgeois salons and the confidence of the czar.

It was here in April 1901, at the headquarters of the Imperial Academy of Sciences, that a telegram from Yakutsk sent hearts racing. The region's governor had sent news of the discovery of a whole, well-preserved mammoth frozen in a cliff along the Berezovka River, just above the Arctic Circle.

Persuaded of its potential importance, Nicholas II's finance minister soon freed up 16,300 rubles—a small fortune, enough to buy a mansion in St. Petersburg, later augmented by the academy's president, Grand Duke Constantine, who donated a few thousand rubles of his salary—to dispatch an expedition after the specimen. Leading the team was Otto Herz, a zoologist with the academy, who brought a taxidermist, Eugen Pfizenmayer, and Dmitrii Sevastianov, a young graduate student in geology. Leaving on May 3, the three took the luxurious Siberian express—a train equipped with a piano, saloon, smoking and reading room, and even a car that served as a Russian Orthodox chapel. After a week they arrived in Irkutsk, a mid-Siberian city near Lake Baikal that was the terminus of the main east-west railroad. From Irkutsk, they traveled several days by horse-drawn wagon to the Lena River. By boat they went to the town of Ust'Kut, where they transferred to a steamboat that got them to Yakutsk on June 1. After three weeks of preparations, another steamboat took the team to Tanda on the Aldan River. They covered the final 1,900 miles on horseback through swampy, nearly impassable land plagued by midges and mosquitoes—a brutal journey that took nearly three months to complete. Recalling the tortuous trek, Pfizenmayer wrote in *Siberian Man and Mammoth* (1939): "Like men, the animals avoid this tundra, said to be bottomless; and all life dies out here. The Tunguses and the Yakuts are firmly convinced that the chief *scheitan*—the lord of hell—lives in the nether regions deep under these malignant swamps, and their tales about him tell of such places."

Exhausted, on August 31 the trio at last reached a village called Srednekolymsk on the Kolyma River. There they met a Cossack, Ivan Yavlovski, who had just returned from the site. Noted for their horsemanship, the Cossacks from the steppes of southwestern Russia and the Ukraine had fanned out across Siberia and settled the region after defeating the Mongolians in the sixteenth century. A descendant of those pioneers, Yavlovski would be their guide.

His initial report was sobering. Heavy rains that summer had eroded part of the slope where the mammoth lay, mangling its hind part. Wolves and other scavengers had nibbled away much of the flesh from its head. To try to preserve the rest of the carcass for the expedition, Yavlovski had piled bones and soil on the exposed parts. News of the damage would have been a crushing blow to almost anyone who had trekked for four grueling months to capture a prize, but Herz remained optimistic. Sevastianov, writing the trip off as a loss, would go no farther. Herz and Pfizenmayer continued on without him.

As they prepared for a 200-mile trip by horseback to Mysovaya, the settlement nearest the mammoth, Yavlovski told them the history of the discovery. In the middle of August 1900, he explained, a hunter named Semen Tarabykin was stalking elk when he spotted a mammoth's tusk protruding high up the cliff. Breaking off the chase to investigate an equally valuable commodity that couldn't flee, Tarabykin discovered a second, smaller tusk closer to the river. This one was still attached to a mammoth's head, part of which protruded from the cliff.

Like other Lamuts, an indigenous people near the Sea of Okhotsk, Tarabykin was superstitious about mammoths, believing that excavating one would bring on a fatal illness. Still, he returned the next day with two friends, who helped him retrieve the 166-pound tusk and hack off the second, not quite half as heavy, from the skull. Several weeks later the Lamuts hauled their booty to a nearby town, where they sold the tusks to Yavlovski. Intrigued by their story of the frozen mammoth, Yavlovski accompanied the Lamuts to the site in November. Seeing the beast for himself, he carved chunks of skin from the head and thigh, as well as a piece of stomach, which still contained the beast's last meal. He took this visceral evidence to Srednekolymsk and presented it to the police commissioner, who decided to launch his own investigation. After visiting the site the following month with Yavlovski, the commissioner, impressed, relayed the news to the governor, whose telegram the next spring alerted the researchers in St. Petersburg.

With Yavlovski as their guide, Herz and Pfizenmayer picked their way on horseback through the sodden taiga near the site along the Berezovka River, where they set up camp. They then hurried to the site. Wrote Pfizenmayer:

> Some time before the mammoth body came in view I smelt its anything but pleasant odour—like the smell of a badly kept stable heavily blended with that of offal. Then, round a bend in the path, the towering skull appeared, and we stood at

> the grave of the diluvial monster. The body and limbs still stuck partially in the masses of earth along with which the corpse had been precipitated in a big fall from the bank of ice. The slopes of this bank rose in places almost perpendicularly about the landslide. We stood speechless in front of this evidence of the prehistoric world.

Getting to the carcass was treacherous: it was sticking out of silt in a collapsed section of the cliff sloping toward the river, 100 feet below. Lacing the crumbly black silt, frozen since it had been deposited during the Pleistocene Epoch, were ice veins several feet wide. One misstep could send them tumbling into the river. Reaching the mammoth mound at last, Pfizenmayer took photographs and began shoveling away the protective silt and bones. Carnivores had indeed devoured most of the head's flesh, but they hadn't found an important clue to the mammoth's death: leaves, grass, and buttercups lodged in the spaces between its four broad, flat teeth. The plants matched those later found in the stomach. Digging a few feet deeper, the team found the left foreleg, covered in bristly rust-brown hairs and a soft undercoat; Herz likened its tan color to a camel's summer coat. The bent foreleg prompted him to speculate on the nature of the creature's death: "It is evident that the mammoth tried to crawl out of the pit or crevice into which he probably fell, but apparently was so badly injured by the fall that he could not free himself." The right foreleg was propped up in a way that suggested that the animal had used it for support while trying to step with the left foreleg.

Digging deeper still, the pair came to the left hind leg, on which they discerned bundles of muscles. As they brushed off the silt with their bare hands, the stench from the carcass grew stronger, limiting their work to minute-long snatches in between which they rushed away, gasping for air. "A thorough washing failed to remove the horrible smell from our hands," Herz wrote. (That very smell makes mammoth meat prize bait for fox traps in Siberia.) Nevertheless, Herz knew the remains were an important scientific discovery, and he paid Yavlovski a thousand-ruble government bounty—which, Pfizenmayer later claimed, the Cossack gambled away that evening. "Was this the luck of the game, or was it another misfortune to be laid at the door of our mammoth?" Pfizenmayer wrote.

If the mammoth was going to be hauled back to St. Petersburg, the scientists were looking at weeks of butchering. Herz hired Lamut workers to build him a cabin with a fireplace. (A photograph on display at the Zoological Museum in St. Petersburg shows Herz and Pfizenmayer, in fur coats in front of the cabin, gripping a 15-foot pole with a dark flag emblazoned with a cartoon mammoth.) The Lamuts also erected a wooden shelter over the mammoth to protect it from rain and snow; they lit a fire inside to prevent the carcass from freezing in the plummeting temperatures of the Siberian autumn nights.

Over the next few weeks the researchers dismembered the beast, sprinkling each piece with alum and salt to preserve it, and transported the pieces to their winter hut. "The most devoted mother could not carry her child more carefully than I carried these fragments of antediluvian fauna," Herz wrote. Some parts that had remained frozen looked appetizing. Herz describes the flesh under the mammoth's shoulder with a flair that could only have come from subsisting for nearly a month on the taiga, far from the comforts of St. Petersburg. The dark red marbled meat, he wrote, looked like fresh beef: "We wondered for some time whether we should not taste it, but no one would venture to take it into his mouth." The scientists stuck with their horseflesh, feeding bits of mammoth to the dogs, which lapped it up.

By early October the stench had lessened—or, Herz thought, maybe they had grown accustomed to it. On the seventh they reached the bottom of the mammoth's hind and found an item for which mammoth DNA hunters would have paid a king's ransom: the beast's squashed, erect penis, nearly three feet long and, flattened, eight inches in diameter. A possible trove of DNA for a cloning experiment if found a century later, the organ and surrounding flesh, even the testicles nestled inside the abdomen, were frozen solid. At last, on the tenth of October, they hauled from the pit a 470-pound piece of abdominal skin. Underneath lay the mammoth's thick, foot-long, bristly tail. The excavation was finished.

Over the next four days they loaded the mammoth parts onto ten sledges pulled by Yakutian horses (a squat breed accustomed to the Siberian winters) and departed the next day, October 15. "Our next problem was to get the sledges with their valuable cargo over the 3,700 miles, across two mountain ranges, from Kolymsk to Verkhoyansk, on to Yakutsk, and thence to the railway at Irkutsk; and all this in the bitter Arctic winter, which daily grew more severe," wrote Pfizenmayer. In the Verkhoyansk mountains, the temperatures plummeted to –67 degrees Fahrenheit. On Christmas Eve they arrived in Yakutsk, where they gorged on food and champagne before heading to Irkutsk by troika, a three-horse sled. A refrigerated train car took the mammoth to St. Petersburg. Their epic expedition ended on February 18, 1902—291 days after it had begun.

The mammoth was taken to the Zoological Museum, where eight days later Czar Nicholas II and his wife, Empress Alexandra, inspected the find. Pfizenmayer wrote:

> The body was reeking with the smell, anything but pleasant, that I have already described; and the whole hall, steam-heated, was full of it. The curator was explaining things to the Tsar, who listened with interest. Not so his wife, who was standing with her handkerchief to her nose . . . soon she broke in with a question: "Haven't you something else interesting to show me in this museum, as far away from this as possible?"

Two years after its discovery, the stuffed Berezovka mammoth, with a reconstructed head and other body parts, was displayed in its death position at the Zoological Museum, where it remains along with its reassembled skeleton. Publishing a photograph of the exhibit, the November 28, 1903 issue of *Scientific American* gushed, "Though many fossil remains of mammoths have been found, and other preserved bodies of mammoths seen, no body so complete as this one has ever before been brought home to civilization." Among other things, the 44,000-year-old specimen revealed that mammoths had four toes (modern elephants have five); that they had a flap of skin protecting their anus from the cold; and that their tusks spiraled toward, not away, from each other. Herz concluded that the Berezovka mammoth had fallen in a crevice. Another possibility was suggested later by the Russian geologist and mammoth hunter I. P. Tolmachoff: the mammoth had become mired in soupy ground in its pasture and suffocated to death. Proving that it had suffocated, he wrote, was the erect penis—"a condition inexplicable any other way."

As much as the Berezovka mammoth was a gift to science, it only reinforced Lamut superstitions that excavating one of the frozen monsters, like entering an ancient Egyptian tomb, carried a curse. Returning to the region seven years after the expedition, Pfizenmayer learned that Yavlovski had gone insane and died. The police commissioner who had led an investigation of the site died two days before receiving a medal from the czar to honor his role in the discovery. The Lamuts were not surprised to learn that Herz, too, had died two years after the expedition, and they feared that Pfizenmayer's apparent good health would not last.

If there is a place where the ghosts of mammoths walk the land, it must be the Berelekh mammoth cemetery. For 20 years Nikolai Vereshchagin had heard field reports of great piles of bones on the Berelekh (Yakutian for "wolf") River in northeastern Yakutia. "The urge to visit the mysterious place preyed on my mind," he recalled. At last he raised the money to gather a team of specialists to travel there in the summer of 1970.

Invited by specialists with the Geology Institute in Yakutsk, Vereshchagin flew to Chokurdakh, a tiny outpost on the Indigirka River. From there they went by helicopter to a camp set up by geologists in a meadow skirted by larch trees. In the 11,500 years since the Great Ice Age relinquished its grip, a dry, grassy steppe that once stretched 300 miles north of today's Arctic coastline had been transformed into a land of a million lakes. Increased precipitation fed these lakes, as did warmer summers, which melted ice in the permafrost. Erosion over the last hundred centuries has eaten away the top 20 to 30 feet of the former Pleistocene plain, leaving behind only scattered mounds, called *yedoma*. These humps of the former steppe are the burial mounds of Ice Age animals. Rivers eat away the *yedoma*, shearing away sediment and liberating bones every summer.

Vereshchagin had never seen *yedoma* mounds before, the

bones of ancient mammoths, horses, and reindeer protruding from dark, loamy sediment. He did not know what to expect at Berelekh. The research party hiked along the river for a half mile, hugging a thin strip of mucky shore between the river and a *yedoma* cliff climbing 60 feet high in places. As they slogged around a bend in the river, Vereshchagin was stunned. "There were bones everywhere," he said. Some were rusty red, most a charcoal gray. Bones protruded from about halfway up the cliff. Bones littered the beach. Bones formed a spit jutting into the river. "We kept silent, struck dumb by the spectacle." There had to be thousands of bones too thick and too long to belong to reindeer or horses. They were the jumbled skeletons of dozens of mammoths.

One group of scientists began counting and sorting the bones. Another set up a gas pump on the river and, using fire hoses, blasted water at the cliff to melt deeper into the *yedoma*. The jets dug several feet deep along a long stretch of riverbank, peeling away the sediment, rich and black like potting soil. The pounding water revealed many more mammoth bones as well as golden and bronze hairs. The water obliterated the configuration of the skeletons, which may have offered clues to how the mammoths had died. But it was impossible to follow the excavation techniques used in temperate areas. "The heaviest pick rebounds from the frozen soil as [forcefully as] from rock," said Vereshchagin.

That summer and the next his group tallied nearly 9,000 bones, together weighing 10 tons. They found a few dozen bones from horses, reindeer, and bison and a single fragment of jaw from a cave lion. But nine out of every ten bones belonged to mammoths, 156 individuals in all. Many were young males, and most of the adults were undersized. In the Indigirka floodplain, conditions near the end of the Pleistocene Epoch may have deteriorated to such an extent that the mammoths' habitat had fragmented into island-like patches, reducing their food supply and stunting their growth.

How did all these animals die? The first summer the scientists found evidence of a Paleolithic camp several hundred feet downstream from the cemetery: stone tools and chips of black and greenish argillite, perhaps flaked from spear points. Then, in the summer of 1971, Vereshchagin stubbed his toe on what appeared to be a mammoth rib sticking up from the sediment along the riverbank. Digging it out, he found an ivory rod carved from a split mammoth tusk. He speculated that Stone Age people attached the rod to a wooden pole, making a spear that could pierce the rubbery hide of the mammoths. Archaeologists from Yakutsk, led by Yuri Mochanov, later uncovered more argillite chips and pendants made from nephrite, a pale green jade. This is the only Paleolithic site ever found above the Arctic Circle.

"For a long time it was thought that people had left this bone cemetery behind," says Gennady Baryshnikov, who as a young zoologist took part in the expedition. Maybe the Stone Age hunters ambushed mammoths, the thinking was, or drove them off a cliff or onto the thin ice of a lake. But that struck Baryshnikov as unlikely. At the campsite were many bones of small game such as hares and grouse. As a general rule, says Baryshnikov, "people ate rabbits, not mammoths." Mammoths, he argued, posed too great a danger to the hunters. An ambush might have required getting in close and spearing a mammoth repeatedly; one misstep could mean getting trampled to death.

Baryshnikov and Vereshchagin had a different idea of what killed the Berelekh mammoths. Packed around the bones in the *yedoma* were scales of cyprinid fish such as the Siberian roach and the decayed remnants of streambed mosses and willows. The most likely scenario, they said, is that these mammoths did not die all at once. Each spring during the late Pleistocene, a few inexperienced young mammoths probably ventured onto the Berelekh's thinning ice, broke through, and plunged into the frigid water, where they drowned and were swept downstream before washing up in an oxbow lake off a bend in the river. Or perhaps they foundered in spring ice floes or in summer floods. The carcasses would rot in the summer, attracting wolves and other scavengers. The hunters too were scavengers, they argued, harvesting tusks for the ivory.

~ ~ ~

As exciting as the Berelekh cemetery was, the discovery of Dima seven years later marked the pinnacle of Vereshchagin's career. Using techniques unimaginable when the Adams and Berezovka mammoths were discovered, Vereshchagin's Russian experts submitted the baby's tissues to a battery of tests, as did two American teams that had received samples from Vereshchagin. These efforts would prove that individual proteins could be recovered from an Ice Age creature.

The leader of one team was the late Allan C. Wilson of the University of California, Berkeley. A New Zealander by birth, Wilson was intrigued by the demise of the moa, a giant bird that apparently was hunted to extinction by indigenous New Zealanders centuries ago. His team hoped to compare mammoth proteins to those of modern elephants and thus learn more about when the species diverged from a common ancestor. Wilson had struck out on two previous attempts with other mammoth remains, finding that his samples were too badly degraded.

Dima's proteins were in pretty bad shape too, Wilson discovered. He couldn't seem to pry them from the tissue: while most undamaged proteins dissolve easily in water, Dima's were insoluble. So he took a different tack, grinding up some of the muscle and injecting it into rabbits, which made antibodies against the foreign proteins. The team culled these antibodies from the rabbit blood and found that they attacked one of the most common proteins found in any mammal: albumin, the main component of blood serum. The antibodies attacked both mammoth and elephant albumin with gusto, further evidence of the close relationship between the extinct species and its modern cousin. Wilson's group later used the same technique to nab mammoth collagen, a protein that forms fibers in bone, ligament, and other connective tissue.

Back in Leningrad, Vereshchagin's team was laying plans for an experiment that, if successful, would trump the American efforts and live up to the omnipresent slogan blared in meter-high block letters on the rooftops of Soviet research institutes: *Slava Sovetskoy Nauke!*—Glory to Soviet Science! What could be more glorious, after all, than cloning baby Dima?

— — —

The man Vereshchagin tapped to lead the cloning effort was Viktor Mikhelson of Leningrad's Institute of Cytology, one of the Soviet Union's top cell biologists. His plan was to search the best-preserved tissues of Dima for cells that could be resuscitated. He would borrow a standard technique used to clone frogs from adult frog cells—an approach nearly identical to the one scientists would rely on nearly two decades later for cloning mammals. If Mikhelson's team found decent cells, it would use a tiny pipette to suck out the nucleus of a live elephant cell, most likely an egg. The cell's inner sanctum, the nucleus holds nearly all of an individual's DNA; it's cordoned off to all but a handful of privileged proteins that maintain or copy the genetic code. Removing a cell's nucleus is like removing a person's brain.

To restore purpose to the enucleated elephant egg, Mikhelson's team would endow it with a nucleus from Dima. They would then zap the egg with electricity to force it to divide and grow. Any embryo that developed would be implanted in an elephant's womb. After about 20 months—the gestation time of an elephant fetus—the surrogate mother, the researchers hoped, would give birth to a mammoth; but success would mean blazing a trail into a whole new realm of science. Clones were by no means science fiction: identical twins are clones of each other, and frogs can be cloned on demand. But no one in 1980 had ever cloned a mammal—let alone an extinct one.

Mikhelson combed the freshest tissues from Dima—muscle, fat from beneath the skin, blood vessels, and blood—for usable cells, dissolving the tissues in a range of solutions for nourishing living cells. After several months, however, he failed to coax a single cell to start working again.

But mammoth DNA did not let Mikhelson down entirely. Since his failure to bring Dima back to life, he and several other researchers have used DNA to help sort out the mammoth family tree. In the mid–1990s, Mikhelson teamed up with Tomowo Ozawa of Nagoya University in Japan to compare DNA from seven species in the Tethytheria, a group of mammals whose ancestors lived some 250 million years ago near the long-gone Sea of Tethys. (The Mediterranean is about all that's left of it.) Among the Tethytheria, protein analyses suggested that mammoths are equally related to the two other proboscideans, the African and Asian elephants, while the two elephant species themselves are more distant relatives.

To probe kinship as written in the genetic code, Mikhelson used the darling of molecular biology, the polymerase chain reaction (PCR) to pluck fragments of mitochondrial DNA from Dima. In PCR analysis, an enzyme makes millions of copies of a target sequence of DNA. Even the tiniest amount of DNA can be amplified by PCR above the threshold of detection for machines that read the sequences of nucleotides, the building blocks of DNA. Mikhelson unleashed PCR on the choicest cut of pickled Dima: muscle from the inner thigh of its right hind leg. He pried from the 40,000-year-old tissue a sequence of 1,005 nucleotides from the gene coding for an iron-laden protein called cytochrome b. This gene is used widely for resolving kinship among species because mutations occur along its nucleotides at predictable intervals.

Ozawa's team extracted the same stretch of cytochrome b nucleotides from six other Tethytheria. In addition to elephants, the researchers compared the mammoth to more distant relatives: the manatee, the dugong, and the extinct Steller's sea cow. These three sirenians may look nothing like mammoths, but subtle characteristics betray a kinship: sirenians and mammoths have similarly constructed hearts and mammary glands that lie between their forelegs, higher than in other mammals. The sixth Tethytheria in the analysis was the rodent-like hyrax, a herbivore found in tropical Africa that has an even more obscure connection to mammoths: similarly shaped ear and leg bones.

The researchers assembled a molecular family tree based on the cytochrome b sequences. The hyrax split off earliest, then the three sirenians diverged quickly. Among the proboscidians, the DNA suggested that the African elephant arose first, around 5 million years ago. That date accords with the oldest known African elephant fossils. Then, about 4 million years ago, the Asian elephant and the mammoth came into their own. But they didn't stray far genetically. "Our work shows that there's very little difference between the mammoth and the Asian elephant," says Mikhelson. Several DNA studies concur, although others have placed the mammoth closer to the African elephant.

Despite his recent success, Mikhelson is reluctant to talk about his naïve hopes in the early 1980s for resurrecting the mammoth. "It was a hopeless task," he says now. Two decades after this doomed effort, he doubts that mammalian cells could survive for millennia—even if kept in a deep freeze and cocooned in a protective environment from the outside world. The landmark achievement reported in 1997 of the first mammal cloned from an adult cell—the sheep Dolly—didn't change Mikhelson's opinion. For now, he maintains, dead cells are a dead end.

But others insist he's wrong. Across the Sea of Japan from eastern Siberia, the reproductive biologist Kazufumi Goto has coaxed life from death.

FIRST DESIGN
THE KOBE STEAK

The rolling hills of Kagoshima and its neighboring prefectures on Japan's southern main island, Kyushu, are beef country. The industry boomed in the early 1980s, when a thriving economy gave many Japanese the means to develop a taste for Western foods, including fine steak. Enlisted as a foot soldier in his country's pursuit of the ultimate steak was Kazufumi Goto.

As a child growing up on Kyushu in the 1960s, Goto developed an intense curiosity about how babies are made, from that transcendent moment when a sperm penetrates an egg and life is conceived until a child is born nine months later. His appetite to learn more went unfulfilled for years, because it was taboo in the highly conservative Japanese society at the time to teach children about such matters before high school.

His fascination with reproduction continued into adulthood. After Goto earned master's and doctorate degrees in animal science at the University of Arkansas, Kagoshima University hired him as an assistant professor in 1983. He began his career studying how to get roosters to produce more semen, but he soon realized that it would be easier to obtain funding from the Japanese government if he got involved in research on the university's 600-head herd of cattle. Goto's plan was to use artificial insemination to develop beef with better marbling—tender, more flavorful meat that could be cut with a butter knife. "We have to compete with American beef," says Goto, a skinny man with bristly hair and a perpetual smile.

His assignment was to design an even tastier steak. Obtaining sperm from bulls with a high percentage of body fat (a measure of marbling) for his breeding experiments was easy. All he had to do was hook them up to an artificial vagina, a tube warmed with hot water and dabbed with pheromones, natural compounds excreted by animals to make the opposite sex take notice. "The bull sniffs at the pheromones and gets very excited," Goto says. "It's easy to do, so long as you train the bulls when they're young." But the experiments went slowly. Goto would insert the ejaculated sperm into cows, hope for fertilization, and wait more than nine months to see if a well-marbled calf was born. "That was a waste of time," he says. "We could only get one good baby every year."

To speed things up, Goto turned to the budding field of in vitro fertilization. In this approach, sperm and egg are united in a test tube and, if conception occurs, the fertilized egg is implanted in a surrogate mother. For the eggs Goto went to a slaughterhouse, where he could grab fresh ovaries from well-marbled cows. "I could see the quality of meat and choose ovaries from the best cows," he says. He could take dozens of immature eggs from a single prime cow, put them in a biochemical bath that prodded them into maturing, mix the ripened eggs with the stud's sperm, and implant the ones that fertilized into many surrogate mothers at once. The surrogate's marbling had no bearing on the calf's. Goto's lab was the first in Japan to develop a technique for artificially maturing cow eggs and fertilizing them successfully.

While these experiments did provide Kagoshima's farmers with a better stock of beef cattle, they failed to quench Goto's desire to do cutting-edge science, and his curiosity pushed him to dabble in the lab after-hours. One day in 1986, while peering through a microscope at sperm swarming over an egg, he saw something extraordinary. "Once a sperm attaches to the egg's membrane, it stops moving," he says. After the egg engulfs the sperm, its whiplike tail detaches from the head, taking with it the mitochondria, or energy source, that powers the sperm. Before fertilization, the sperm head (a crucible for the father's DNA) lacks the energy to fuel the biochemical reactions that keep a cell alive. Technically, it's dead.

Staring at the stationary sperm, Goto suddenly thought: if sperm dead for seconds can fertilize an egg, why couldn't sperm dead for years or even centuries? Amazed by the implication, he started an experiment that would challenge the meaning of life. But first he had to kill the sperm—no easy task. Goto knew that sperm could survive for decades in a freezer so long as they were bathed in chemicals that protected them against the ravages of freezing. Cryoprotectants commonly used in reproductive experiments include glycerol and ethylene glycol—automobile antifreeze. "If you freeze sperm without cryoprotectant, they will die," Goto says. Ice crystals forming inside the cell rip the sperm's fragile membrane, and its internal chemistry grinds to a halt.

He instructed a new graduate student in his lab to freeze

and thaw the semen of mice—the most common lab animal—without cryoprotectant until the sperm were dead. Since immobile sperm may only be playing possum, Goto ran the beat-up sperm through two more procedures to eliminate any possibility that they were alive. First he and his student put a dye into the solution containing the thawed semen; if the sperm membranes were broken, the dye would easily infiltrate the sperm head. That was indeed the case. Next they checked to see if the sperm mitochondria were still making ATP, the biological molecule used by cells for fuel. They found nary a trace of ATP production. The rough handling posed little threat to the sperm's genes. "We could freeze and thaw twenty times, and the sperm DNA would never break," Goto says.

But when they tried a new technique—intracytoplasmic sperm injection—to inject the DNA-packed dead sperm heads into mouse eggs using a slender glass needle, the experiments failed time after time—a few thousand times, that is, over two years. The membranes of the mouse eggs were so fragile that they would rupture whenever the researchers punctured them to inject a sperm head. "We didn't know at the time that the mouse was a difficult model for this kind of experiment," Goto says.

He retreated to more familiar turf, instructing the student in 1989 to try the same experiment using bull semen and cow eggs, which have tougher membranes. After injecting the dead sperm into hundreds of eggs, the student—who had never seen a sperm or egg before coming to Goto's lab—one day yelled, "Hey Goto! We've got cleavage!" What he meant was that fertilization had occurred: looking into a microscope, he had seen an egg that had cleaved several times to form a blastocyst, an early embryo. The men wasted no time implanting the blastocyst into a cow. A calf was born. Repeating the experiment many times, they produced four more calves.

Racing against an American embryologist who was also experimenting with intracytoplasmic sperm injection in cattle, Goto published his work quickly in an obscure British journal, *The Veterinary Record,* in 1990. For two years few people outside Japan knew what had been accomplished. Besides reporting the first success using intracytoplasmic sperm injection in cows, Goto's team had got the first live births of any animal using dead sperm. As they pointed out in their paper's conclusion, "The result raises the question of what is meant by the death of spermatozoa."

Goto pondered that question for months, until a journalist helped him find the answer that set his career on a new path. In 1992, at a meeting of the International Embryo Transplant Society in Denver, Goto trotted out his team's findings. The presentation wowed the audience, which included Robert Cooke of *Newsday,* a Long Island, New York, newspaper. The reporter then interviewed Goto about the meaning of his work. Besides staving off extinction for species that cannot breed on their own, Goto ventured, the technique might be used to try to produce a child from the 5,000-year-old "Ice Man" discovered frozen in a glacier in the Alps on the Italian-Austrian border. He acknowledged that such an experiment would raise ethical howls, and he did not advocate trying it. But the conversation turned to an idea that Goto considered thrilling and morally defensible: resurrecting an extinct animal like the mammoth.

All Goto had to do was lay his hands on a frozen carcass and its precious sperm.

Numerous scenarios have been proposed to explain how Ice Age animals occasionally turn up in Siberia in a remarkable state of preservation. One of the more fantastic ideas appeared in 1913 in *A Journey to the Earth's Interior,* by Marshall Gardner, who claimed that the earth was a hollow sphere, with slits at the poles, enveloping an inner sun. He maintained that the inner surface also supported life, including the mammoth. Frozen mammoths in Siberia were simply the remains of ill-fated individuals that had wandered through the hole in the North Pole.

As entertaining as Gardner's idea was, most people realized that mammoths were once part of our world and that it required rare circumstances for a carcass to freeze and last for millennia. The first clue to this mysterious process came in the middle of the nineteenth century, when the explorer Alexander Theodor von Middendorff found a whale's carcass buried under silt on the shore of the Sea of Okhotsk. It was unclear how long the whale had been lying there—it may have been years, decades, or longer—but its fat was still edible. Middendorff speculated that mammoths that foundered in muddy pits or in marshes may have been preserved this way as well.

Goto's recipe for bringing back the mammoth—by using sperm to breed a hybrid mammoth-elephant, which after three generations would be nearly 100 percent mammoth—began with well-preserved mammoth sperm. That in turn required a serendipitous death. "The body should be frozen quickly and protected from the elements," says Nikolai Vereshchagin. A quick, hard freeze is necessary to preserve the tissues before decomposition sets in. (Even in the harshest conditions, however, it would take hours for the several-ton body of a mammoth to freeze solid, allowing ample time for bacteria to begin to putrefy the innermost meat.) Besides staving off bacteria, rapid freezing would ensure that the liquid inside the cells formed tinier ice crystals. The smaller the crystals, the less likely they would be to rip vital organelles or burst a cell apart. Within hours of freezing, the carcass must be covered in snow or otherwise shielded from scavengers as well as unfavorable conditions: warm temperatures that could spoil it, winds and dry air that could dehydrate it.

Only a few manners of death might conceivably preserve a mammoth for millennia. One is to be encased in silt or water, then quickly frozen. The victim may have become mired then sank, as Dima appears to have done. Or the animal could have been swept away by a river of icy mud during the spring flooding, when melt water that can't seep into the permanently frozen ground sweeps into the rivers. Another possibility is that a mammoth broke through the icy crust covering either small ponds or treacherous permafrost potholes in labyrinthine formations of aufeis, sheets of ice formed when ground water rising to the surface freezes in winter.

A second measure of serendipity must occur in the present. "We must find the body when it's still frozen," says Vereshchagin. Decay resumes the moment dead tissue begins to thaw.

Hoping to get lucky, Goto wrote to several Russian scientists; they never replied. He asked the Russian embassy in Tokyo, but they asked Goto if he was really after mammoth ivory. "I had no success at all," he says.

In 1995, however, a university colleague put Goto in touch with a Japanese businessman who had good contacts in Russia. Kazutoshi Kobayashi, the president of a trading firm called Field Co., Ltd., was no stranger to bizarre ideas; he was dealing in feces-eating flies, imported from a secret Russian laboratory, whose larvae digest excrement from farm animals. Goto made an appointment to meet Kobayashi at his headquarters in nearby Miyazaki, on the southern coast of Kyushu.

A former resort town—Japanese honeymooners and sun lovers flocked to its beaches and palm-lined boulevards after World War II—Miyazaki today is a pale reminder of that heyday, having lost many of its tourists to more exotic destinations like Hawaii.

Kobayashi remembers vividly his first meeting with Goto, in February 1996. "He talked like a child about his dream to make a mammoth, but he was serious," he says. "My friends always thought I was the most foolish person in the world, but there before me was someone even more foolish." Still, Kobayashi recognized a kindred spirit: "When I was young, I didn't have any foolish dreams, I just cared about money. Then I developed a kind of disease: I wanted to do something others considered impossible. I told Goto we would realize the dream together." When Kobayashi told his staff about the new venture, they thought he had gone mad. He tried to reassure them. "I told them, 'There is so much scary news out there. It's good to have at least one good news story.' I told them, 'We are losing the romantic things in the world.'" A few weeks later, Kobayashi phoned Goto and proposed a trip to Siberia to excavate a frozen woolly mammoth. He told Goto not to worry about money, to concentrate on the science instead.

Goto's first priority was to find frozen mammoth sperm with which they could inseminate a close relative of the mammoth—the Asian elephant—to produce a hybrid. He thought it would be easier to find usable mammoth sperm—which, like all germ cells, has only half the full complement of DNA—than it would be to find an intact cell from elsewhere in a frozen carcass. (This strategy had been anticipated a decade earlier. A report in *Technology Review,* a magazine published by the Massachusetts Institute of Technology, described how a veterinary researcher at Siberia's University of

Irkutsk obtained egg cells from a frozen woolly mammoth. The scientist fused the eggs with sperm from an Asian elephant and, through "retrobreeding," produced a living hybrid. Dubbed a mammontelephas, the hybrid was supposed to earn its keep by hauling timber for loggers or by using its trunk to pull disabled trucks out of snowdrifts on the trans-Siberian highway. The *Chicago Tribune* and other newspapers that picked up the story failed to notice that it was dated April 1, 1984: April Fool's Day.)

Goto knew that finding intact mammoth eggs was highly unlikely. Researchers had only been able to freeze the eggs of four species—hamsters, mice, cattle, and humans—and thaw them intact. But finding well-preserved frozen mammoth sperm, while a long shot, seemed possible. Adult males would be a problem. Their testicles, nestled inside their abdomens, would almost certainly have contained sperm that had degraded in the residual warmth of the corpse. However, young males or dwarf males that died and froze quickly might have shed body heat so rapidly that the sperm would have been preserved.

Even then, it was unclear how well mammoth sperm would keep on ice in an environment bathed in DNA-damaging radiation streaming from the sun, the cosmos, and radioactive elements. Indeed, the fragility of sperm varies enormously across species. Human sperm, for example, have been kept on ice, intact, for 50 years, and some experts believe they can last for millennia. The sperm of many other species, however, are much more fragile: monkey sperm and pig sperm, for example, spoil quickly. Scientists are only beginning to get a handle on why sperm from certain species are so fragile. For example, scientists at the Institute of Zoology in London have found that while the sperm of kangaroos and wallabies survive freezing and the initial thawing, their membranes fall apart when the broth they're in is raised to room temperature—a phenomenon apparently unique to marsupials. Even closely related species exhibit a range of tolerance. Disturbingly for mammoth enthusiasts, African elephant sperm keep relatively well, while those of Asian elephants—a closer relative, perhaps—fare poorly. The good news is that Asian elephant sperm appear to be relatively hardened to radiation. In experiments, sperm bombarded with a dose of radiation (the primary cause of DNA mutations) equivalent to that which might be received naturally over 3,000 years had no apparent ill effect on the sperm's ability to function.

Still, Goto knew that even if he could find good mammoth sperm, his team would face daunting technical challenges. The first step would be to see if the sperm's chromosomes were intact. To do that, the researchers would inject the sperm DNA into a mouse egg and watch the tightly coiled molecule unwind in the unfamiliar cellular milieu. Unbound, the chromosomes would spread out over several hours. A failure to fan out would mean trouble. If the mammoth sperm DNA was found to be intact, Goto predicted, "it should be no problem to get offspring."

He would inject only mammoth sperm with an x chromosome into elephant eggs, thus ensuring that any hybrid born would be female (mammoth sperm with a y chromosome would guarantee a male baby). When the hybrid reached reproductive age, it would be impregnated with mammoth sperm. Most hybrids of two species are sterile, but there are exceptions. One factor is the number of chromosomes of the mated species. While mules—the offspring of horses (with 64 chromosomes) and donkeys (62)—are sterile, river buffalo have two more chromosomes than swamp buffalo, but the two species can reproduce with each other over several generations. Thus, even if mammoths had one pair of chromosomes more than elephants, the offspring of the two species might still be able to reproduce. Asian and African elephants are farther apart genetically, it appears, than either is to the mammoth; the few attempts to mate the two elephant species have met with disappointment, yielding only a short-lived hybrid. But Goto wouldn't know how a mammoth-elephant hybrid would fare until he tried to make one. And for a crack at this experiment, he would need Kobayashi and his Russian connections.

Hardly your typical middle-aged Japanese businessman, Kobayashi prefers a turtleneck to a suit and blue suede boots to wingtips. He started out in advertising, establishing Field Co. in 1981 with fewer than a dozen employees; they designed everything from office space to logos for the Japanese air force. But in December 1993 Kobayashi, who enjoys hunting and horseback riding, embarked on a new adventure: he flew to Khabarovsk, in Russia's Far East, to look for scientists who might be willing to license inventions.

"Ten out of ten friends opposed the idea of doing business in Russia," Kobayashi recalls. And they were nearly proved right. During six years of trips to the chaotic land that emerged after the USSR collapsed, Kobayashi endured suspicions that he was a spy, survived an emergency landing in Siberia, and overcame the reluctance of Japanese banks to finance deals in Russia. But the tribulations were worth it. His office wall is covered with the pictures of dozens of scientists with whom he struck deals on everything from a new kind of metal-cutting blowtorch to oxygen generators used aboard the Mir space station to the feces-loving flies. After licensing the technologies from the Russian institutes, Field Co. markets them to major Japanese companies.

Mammoth meat, however, was one commodity in which Kobayashi had never traded. But he did have ties to Yakutia, a Siberian province vaster than modern India. A few months before Goto arrived on Field Co.'s doorstep, the firm had finished blueprints for the interior of a new sports center near the province's capital, Yakutsk. The center's director put Kobayashi in touch with the veteran mammoth hunter Pyotr Lazarev, who in 1991 had opened a Mammoth Museum in Yakutsk.

A geologist by training, Lazarev had made expeditions in Yakutia with Vereshchagin and others to recover mammoth bones and frozen carcasses. The Japanese needed a professional who knew where to look and how to excavate in the permafrost; Lazarev seemed to fit the bill.

Lazarev knew the irreparable damage to frozen mammoths that amateurs might inflict. In the fall of 1986, a reindeer breeder from a collective farm, the Dawn of Taimyr, discovered the body of a mammoth frozen in a bank of the Yekaryauyakha River in northern Siberia. The carcass was decapitated, with the head frozen in the shallows. The breeder sawed off the valuable tusks and contacted a regional official, who then called in scientists who had never excavated a mammoth. In a major blunder, they thawed it out and left it lying in the sun on a warm day, then went to the nearest village to wait for experienced mammoth researchers to finish the job. Hours before the pros arrived, however, the mammoth carcass tumbled onto the river ice and, along with the head, was swept by floodwaters into the gulf. Lazarev was eager to avoid a similar incident, so he gladly agreed to help the Japanese.

Barely six months after Kobayashi and Goto first met, Lazarev invited Goto and a colleague to visit Yakutsk in August 1996 for a preliminary excavation sponsored by Field

Co. When the Japanese arrived, they got a crash course in Russia's finances—the crumbling apartment buildings, the frequent power failures, the grandmothers selling cigarettes on street corners to scrounge enough rubles to buy a loaf of bread. "I thought I had traveled back thirty years in Japan, to when I was a child," Goto says. But the people he met were clearly resourceful. He wondered how well his own countrymen would endure such hardships. "If we brought Japanese people to Russia, maybe they would not survive," he says. "Some Japanese people never wash dishes or clean houses. They don't have enough life experiences."

Lazarev took the scientists by motorboat along the Lena River west of Yakutsk, where they spent three days scouring the riverbank and digging in the Pleistocene sediments for frozen carcasses. "I realized the chance of finding tissue in only a few days was very small," Goto says. His hunch was right. They found mammoth tusks and bones, but no meat.

Although the excavation was a bust, it stoked enthusiasm for a more ambitious project. Kobayashi brought Lazarev to Japan in February 1997 to plan a major expedition for that summer. He also founded the Creation of Mammoth Association to publicize and run the project, after which he signed an agreement with the Sakha Republic, which oversees Yakutia, to share custody of any mammoth the Japanese researchers brought to life. For the association's logo, Field Co. designed a cartoon mammoth with an oversize bright red heart.

Goto, meanwhile, didn't want to share the burden of the science alone, so he invited Akira Iritani, whom he had known for almost a decade, to join the team. Iritani came with impeccable credentials. A highly regarded reproductive biologist and chairman of the Department of Genetic Engineering at Kinki University near Osaka, Japan, Iritani and his team had achieved the world's first birth of a mammal (a rabbit) using intracytoplasmic sperm injection in 1986. This technique has since become indispensable in human fertilization clinics, producing as many as 25,000 babies worldwide each year.

A few months before Iritani joined Goto, Ian Wilmut's group in Scotland made a huge splash with the news that they had succeeded in cloning the sheep Dolly from an adult cell. Encouraged, Iritani eagerly joined the sperm hunters, but he worried about how long it would take for the project to succeed. If a mammoth sperm and an elephant egg were united to create an embryo that, in turn, took to a surrogate mother, a hybrid (let's call it a mammophant) would not be born for another 20 months. If a hybrid were fertile, it should be able to reproduce after 10 or 15 years. If all went flawlessly, it would take at least 35 years, from start to finish, to get a mammophant that was roughly 90 percent mammoth.

Iritani had another idea. If even a small bit of mammoth tissue were found with normal DNA, it would take only one procedure to clone a pure woolly mammoth. He thought the best place to look for cells with intact chromosomes would be in the testicles and ovaries. (While cells elsewhere in the body go through periodic bursts of activity, including division, the Sertoli cells from the testis and the cumulus cells from the ovary are perpetually in quiet mode, never dividing.)

Following the standard cloning procedure, Iritani wanted to fuse a well-preserved mammoth Sertoli or cumulus cell—each of which has a full complement of DNA—with an unfertilized elephant egg stripped of its nucleus. If an embryo developed, he would let it divide to the 32-celled morula stage. Then he planned to send the cells to a lab at Thailand's Mahidol University, where Asian elephant eggs had been successfully fertilized in a test tube. There, the mammoth embryo would be implanted in the womb of an Asian elephant, where it would develop into a fetus that was 100 percent mammoth. "There's nothing unique about elephant eggs that would interfere with the experiment," says Iritani.

This idea gave the Japanese team two routes to scaling their Mount Everest of biology experiments. But with either approach, the risk of failure was high. No one knew whether mammoth sperm could inseminate an elephant egg—the crucial step in creating a hybrid. And the odds of finding a frozen mammoth cell bearing intact DNA were incalculable; Iritani would have to score a scientific coup in finding such DNA before he could try to clone a mammoth. Neither team could claim that its technique stood a better chance of success, so Goto and Iritani enjoyed a playful rivalry over who would make history.

With two possible plans and a stellar crew in place, Kobayashi flew to Yakutsk in June 1997 to grease the bureaucratic wheels for the expedition. He also brought a load of cash to sponsor a preliminary dig by Lazarev's group. But disaster struck on the way to Yakutsk. According to Kobayashi, a bag of his with tens of thousands of dollars inside was stolen in the customs hall in Seoul, Korea. Credit cards weren't accepted at many Russian establishments outside Moscow, and Japanese are accustomed to paying most of their bills in cash. "It was terrible," he says. "I had to go to Russia without any money."

On his arrival, Kobayashi scraped up a few thousand dollars from his Russian associates and sent Lazarev's team on an exploratory mission: a 1,000-kilometer helicopter trek to Chokurdakh, the tiny settlement on the Indigirka River that Vereshchagin had passed through on his way to the Berelekh mammoth cemetery. The team found a chunk of mammoth leg. Although the flesh was not particularly well preserved, the find suggested that Chokurdakh was a suitable spot to return to in August. Kobayashi got written permission from the president and vice president of the Sakha Republic for his group to enter the village, which was usually closed to foreigners (as are so many other towns in the Siberian Arctic). But a week before the Japanese departed for Russia, the mayor of Chokurdakh demanded $70,000 for permission to enter his demesne. Kobayashi balked, negotiations collapsed, and Lazarev began mulling alternatives.

He settled on Duvannyi Yar, Old Russian for "windy cliff," a renowned mammoth site reached via Cherskii. This town high above the Arctic Circle had prospered under the Soviet leader Nikita Khrushchev's attempts to shore up the Soviet economy by extracting gold, timber, and other natural resources from Siberia. Yakutia, for instance, had become one of the richest regions in Russia after the discovery of diamonds there in the late 1960s. Lazarev phoned the ecologist Sergei Zimov, who headed a scientific station in Cherskii. Zimov quickly offered to arrange for a barge to ferry the 30-member expedition up the Kolyma River to the site.

The group soon ran into trouble. On their arrival at Khabarovsk airport on August 10, customs officials seized a satellite phone and a detailed map of the region published by a U.S. government agency. No matter that Kobayashi had secured all the necessary permissions in writing from Sakha officials. "The customs officers just wanted us to pay arbitrary fines," says Iritani. Money changed hands, but the complications cost the group three days. Finally, on August 14, the team reached Yakutsk, where all multiple-story buildings in the dusty capital of Yakutia are built on pilings to anchor them and prevent them from slowly sinking into the permafrost. The Japanese were to meet Lazarev here and fly to Cherskii.

When Kobayashi went to inspect the chartered plane, he discovered a serious problem. "The plane was full of goods for the black market," he says. Furious, he demanded that the pilot unload the plane to make room for the expedition's cargo. The pilot refused. After two days of negotiations, Kobayashi finally paid the charter fee, unwilling to waste any more precious time. As they were boarding, the scientists noticed a few strangers crouched behind the illicit goods. "It was all very suspicious," says Kobayashi.

The scientists were relieved to arrive in Cherskii. The resourceful Zimov had bought a barge and on it built a sleeping shelter, a rudimentary kitchen, and a pair of outhouses. There was even a sauna. "After all the troubles we had," says Kobayashi, "when I saw the houseboat, it looked like a luxury hotel."

Energized by warm, sunny weather and sporting the Creation of Mammoth Association's flag emblazoned with a cuddly mammoth, the group set out on a day's cruise to the 20-kilometer-long Duvannyi Yar cliffs. The adventurers breathed in the fragrance of alpine sage and enjoyed moose stew and white nights, when the sun never sank below the horizon. Birch trees, a brilliant crimson, covered Mount Panteleikha up to the snowline. This land, they thought, once was the home of a mammoth—and might be so again someday.

The expedition anchored off Duvannyi Yar, foraying to and from shore by rubber boat. Leading the way was the research station's little black poodle—not the kind of dog one might expect in Siberia. "He was like a scout, showing us the dangerous places where we shouldn't step," says Iritani. The scientists set to work, scouring the cliff for signs of mammoth tissue, which they would bring back immediately. They would spare no expense to fly a promising specimen by charter from Cherskii to Japan. At home, they planned to use their existing labs—no special mammoth equipment needed—for the revolutionary experiments.

Using picks to probe the cliff face, the researchers homed in on spots with a high density of mammoth bones or swatches of hair. They labored under blue skies in T-shirts, their skin slathered with high-octane bug spray to ward off the swarms of mosquitoes. They scampered up and down the cliffs, some 80 feet high, avoiding the slick curtains of ice that formed in fissures where the ground had cracked in hard freezes. "It was a dangerous place; sometimes sections of the cliff would collapse," says Iritani. On the first day, a bear and later a wolf were spotted eyeing the group from the top of the cliff. From then on, Kobayashi patrolled the excavation with a shotgun. The risk and isolation, he says, made people edgy. Fortunately, neither beast nor landslide claimed a victim.

By the third day, all the expedition had to show was the skeleton of a Pleistocene horse. The weather started to turn worse, and the Russians pressed to return to Cherskii; after heated debate, the Japanese relented. "It was frustrating not to come back with meat," says Kobayashi, particularly after his company had spent 40 million yen, or about $350,000 at the time, for the expedition, only part of which was paid by two Japanese TV networks that filmed the mammoth hunt.

Goto's first attempt at finding mammoth sperm or pristine cells failed. But he and his colleagues vowed to try again the next summer.

RIVER OF BONES

The idea of bringing an extinct species back to life seemed absurd, more like science fiction than reality, when I first saw a newspaper report on it early in 1998. Were Goto and Iritani imposters of the *Weekly World News* ilk? I wondered. Conversations with experts in reproductive biology persuaded me that they were, indeed, respectable scientists. Tracking down Goto at his office at Kagoshima University, I eagerly agreed to join him on his next frozen safari.

I thought I was prepared, but I was wrong: I knew a very different Russia. In the mid-1990s I lived in Rostov-on-Don, the capital of a southwestern region known for its Don Cossacks, fine champagne, and enterprising mafia. Like New York's early in the twentieth century, Rostov's quays along the Don River—a major shipping route from the Russian heartland to Europe and the Middle East via the Black Sea—were rife with mobsters. This had been true even in totalitarian times. Today they spend their money in pricey boutiques along

Rostov's main street, Big Garden Avenue, or gamble it away in casinos, where $100 is the minimum bid at many blackjack tables.

The new Russia mixed uncomfortably with the old. I taught journalism for a year at Rostov State, where professors hoarded bits of chalk and often went months without receiving their salary—less than $100 a month. Many academics held second jobs—driving taxis, for instance, or peddling Nescafé in outdoor markets—and got through the winter by subsisting on vegetables grown at their dachas. For me, Siberia conjured up images of wintry desolation, snow accumulating on abandoned gulags. I had not imagined that the snow accumulated on the bones and flesh of mammoths as well.

I had arranged to meet Goto's group in Cherskii in late August 1999, before winter closed in. Getting there meant flying halfway across Russia on an overnight flight from Moscow to a settlement called Tiksi. As we descended, the Lena Delta glittered in the morning sunshine, a panorama of streams and rivulets wending through bare tundra. Only on the final approach was this stark land marred by the ramshackle town, a sprawl of coffee-colored wooden buildings subsiding into the tundra. Most Russian maps feature Tiksi prominently as a regional capital, with only a handful of other towns depicted for hundreds of miles. As I later learned, many of these "towns" are little more than one- or two-person weather stations.

I had expected to stay in Tiksi only long enough to buy tickets for the weekly flight to Cherskii, but I soon ran into trouble. When the plane landed, the border guards came aboard and checked everyone's passports and other documents to make sure all of us had permission to be in Tiksi. They gravely inspected my foreign passport and escorted me individually into the aqua blue and gold terminal. I passed through a dingy hall with wooden benches and a ticket booth, to a back room. During the Cold War, foreigners were largely denied access to Tiksi and many other towns in the Siberian Arctic, which often had military bases and surveillance outposts to guard against sneak attacks from Alaska. In the 1990s restrictions were eased, and foreigners could visit Arctic towns if they were invited, but the local authorities still insisted on having detailed itineraries. The border guards transcribed my passport information and recorded the answers to stern questions such as *"Shto Vy delaete v Tiksi?"*—"What are you doing in Tiksi?" A reasonable question, and one that I was asking myself just then. With relish I replied that I was just passing through.

After the guards made sure the carbon copy of the information they took down was legible, they allowed me to join the throng milling outside the ticket window. It's impossible to buy a ticket for the Tiksi-Cherskii route outside either town, which makes passing through Tiksi a necessary gamble. But it was a gamble nonetheless, especially in late summer, with people returning to their Arctic homes after brief holidays in Moscow. Especially when the flight from Moscow was in a packed jet and the only other airworthy plane on the runway—the plane most people were taking—was a stubby prop plane. The math seemed not in my favor. The flight connection was like cramming leftovers into a plastic container and discovering, too late, that there's too much food for it to hold. I wondered if I should have taken one of the two alternate routes, either flying into Yakutsk, a tack that Sergei Zimov had discouraged—the border guards there were said to give foreigners continuing on to outlying regions a difficult time—or flying into Magadan and taking a three-day motorboat ride with Zimov up the Kolyma River to Cherskii. Next year, I thought, I would take the scenic route.

My mood darkened considerably when the clerk sold the last three tickets to Cherskii with at least a dozen people pressed against the window ahead of me. Some people asked politely if they could buy a ticket; others begged or cajoled. Whether faced with sweetness or bellicosity, the brawny female clerk's reply was always the same: *"Ne trebuyte ot menya nevozmozhnovo!"*—"Don't ask me to do the impossible!" The only visible piece of electronic equipment, a board meant to display flight times, was not working. I had no clue if the flight was still boarding or had already taken off.

One of the border guards, a guy in his twenties in a smartly pressed olive-green uniform, approached me and asked if there was a problem. I told him there were no more tickets to Cherskii, to which he replied, "We have a nice hotel in Tiksi. You can stay here until the next flight." I thought he was joking in a wry Russian way, but even after a few seconds he didn't show a hint of a smile. Pleasant as seaside Tiksi might have been, I didn't want to be stranded there for a week. At that moment I yearned to be on Zimov's motorboat, making for Cherskii.

But I didn't give up hope, not only because none of the other travelers had abandoned the ticket window, but also because I'd learned that an impossible situation in Russia offers a greater hope of resolution in one's favor than a similar situation in America. After nearly an hour of uncertainty, an administrator emerged and announced that while all the seats had been sold, the plane—its passengers and cargo fully loaded—was still under its weight limit. With the determination and efficiency of someone finishing a distasteful task as quickly as possible, she weighed the rest of us as cargo. I cheerfully paid for the impossible ticket.

Boarding the Antonov jet with a stylish red racing stripe on its side through a cargo ramp in the rear of the craft, it hardly mattered that I could barely make out the people and the cargo sharing the cabin. The crew hadn't switched on the lights, and there were only two pairs of small windows at either end of the plane. Bags and palettes were stacked to the ceiling behind several rows of seats. Was the plane overloaded? Who cared! We were on our way.

A few hours later I arrived at Cherskii. Lugging my bags across the tarmac, I rejoiced to see a tall man with thick wavy blond hair waving to me. It was Sergei Zimov. Thanks to his collaborations with Western scientists that bring in badly needed cash for his research station, Zimov drove one of the more impressive—and one of the few operable—vehicles in Cherskii, a Japanese four-wheel-drive. Cherskii was once a bustling Arctic port that shipped out gold from a major mine 180 miles to the east. Now, all but a few of the dozen giraffe-like cranes on its docks lie dormant. An abandoned wooden *stolovaya,* a cafeteria, was subsiding into the permafrost; not far away, a three-foot-wide metal snowflake, a Christmas decoration on a street lamp, dangled precariously from the pole. Splintered wood and scrap metal littered the shoulders of the unpaved roads, turning the entire town into a junkyard. As poor as the hyperborean outpost of 8,500 souls appeared, however, it had managed to stabilize its economy and was one of the few Arctic towns expected to survive Russia's prolonged malaise. One bright spot is the Lower Kolyma College of Northern Peoples, where Yakuts can take courses, for example, on the business of herding reindeer.

Cherskii was depressingly dark. In a bid to save money, the

town had turned off the electricity to half the buildings after moving the residents into vacant apartments—abandoned by those who had returned to European Russia—in the electrified district near the Kolyma River. A few dozen squatters, betrayed by the eerie candles in the windows, took refuge in the abandoned blocks. They posed little threat to the community, however. Practically the only crimes in the town stem from bar fights or other alcohol-related misadventures. "Nobody locks their car around here," Zimov says. Well, almost nobody owns one, I thought.

We traveled along a dirt road leading out of Cherskii into taiga, forested land where the ground a few inches below the surface is permanently frozen. Zimov weaved around potholes that might have swallowed a Lada, if one of the pint-size Russian cars were to be found here. We passed a bank of mobile, dark green, military radar antennas. Relics of the Cold War, the antennas and their swivel bases were now rusting away. Paranoia about a sneak attack once ran so high that teams of scientists in the 1960s and 1970s fanned out across northeastern Siberia to study whether American tanks, transported from Alaska, would be able to roll across the tundra or get bogged down.

On occasion, Communist party officials feigned indifference to foreigners wishing to visit the Arctic. In 1979 Soviet scientists were granted permission to plan a major conference in Cherskii on the geology of northeastern Siberia. "We invited all the leading Canadian and American scientists," recalls Andrei Sher, an expert on Pleistocene mammals in Moscow. But the party's Central Committee had to give permission for the visas. "Finally, one by one, the visas were all refused," he says. Then, only a few days before the conference was due to begin, the committee did an about-face, giving Sher and the other organizers its blessing to invite several dozen foreign specialists. By then it was too late for most of the invitees, who had made other plans. "It was a very stupid trick by the Soviet government," Sher says.

In a speech in 1988, Mikhail Gorbachev promised that the Arctic would be opened for research, and that December Soviet scientists convened a meeting in St. Petersburg to start planning joint projects. The first few years after the Soviet collapse, research flourished. In the mid-1990s, however, Russia began to regress to its Cold War attitude, placing more and more restrictions on foreigners intending to travel to the Arctic.

Thus I considered myself fortunate to be able to peer behind the icy curtain. Leaving the decrepit military hardware behind us, we bounced around a bend after a few miles and ascended a hill to reach three houses and several outbuildings perched on a bluff. The compound overlooked the Kolyma River and a patchwork of lakes and sedge meadows that ran for dozens of miles to two humpy, snowcapped mountains. It was Zimov's ecological research station, a branch of the Pacific Institute of Geography in Vladivostok, 2,000 miles to the south. As we pulled up to one of the houses, a black poodle ran up to the car, barking. This must have been the intrepid scout that helped the Japanese expedition avoid danger zones at Duvannyi Yar.

Awaiting us inside were Zimov's wife, Galina, and a delicious dinner of moose pilaf. Zimov doesn't particularly care for hunting, but every few months he and one of his staff, another Sergei, shoot a moose to feed the three families at the station. As we ate and talked over a bottle of vodka, Zimov shared his skepticism about the Japanese hopes of finding frozen mammoth tissue. Sure, he said, plenty of mammoth remains are entombed in the permafrost. After 25 years of observation on the taiga and tundra, he and his colleagues had found that the so-called Duvannyi Yar sediments near Cherskii span some 90,000 years and contain roughly 600 skeletons per square kilometer. Assuming that conditions for mammoth preservation were good for only certain periods during the late Pleistocene, Zimov estimates that the average density of the mammoths around Duvannyi Yar during the Great Ice Age was sixty individuals for every hundred square kilometers—a density rivaling that of elephants on the African savanna. How many of those carcasses remain frozen today is impossible to say. The trouble for the Japanese team, Zimov said, is that "frozen mammoths find you, not the opposite." Only a few dozen frozen mammoths have ever been discovered, so "you have to be lucky to find one," he said. "Directed searches have almost no chance of success."

I hoped to prove him wrong. I had come to find a mammoth.

- - -

Zimov had invited Goto's group to explore a wider area beyond Duvannyi Yar that year by helicopter—or by swamp boat, which would not be at the mercy of the low cloud ceiling that often brooded over the floodplain, delaying flights. The cost of hiring the sole helicopter at the Cherskii airport had skyrocketed from $200 to nearly $1,000 per hour that year, so the Japanese asked Lazarev to hire a commercial trawler for the expedition. Lazarev led Goto's group on a hunt into new territory, while Zimov took me to the richest stretch of Pleistocene riverbank: Duvannyi Yar.

The day was overcast and chilly, and I huddled against the wind during the two-hour motorboat ride to the site. On one side of the river, the bank grew steeper and revealed rich, loamy black sediment, rising more than 80 feet in places. At the top of the crumbling cliff, larch trees leaned drunkenly, unable to keep a secure grip in the thin, thawed soil. Zimov cut the engine and drifted closer to shore. About twenty feet from the beach he jumped out of the boat. The water came only halfway up his shins. He grabbed a rope tied to the bow and started plodding parallel to the shore with the slow current.

Stepping out of the boat, I discovered why Zimov was having a difficult time of it; with every step I sank into the muck, which seemed to create a vacuum lock on the rubber of my hip waders. I lumbered as ponderously as a woolly mammoth toward shore. The sucking noises from my boots as the silt relinquished its grip seemed like the lost sound of a mammoth yanking up grass with its trunk and slurping it into its mouth. Each step closer to the quicksand-like beach became harder.

The odor of decaying life was everywhere. So were bones. I could only guess what parts of the animals they represented. But they were all Pleistocene. As I picked up the relics, Zimov called out identifications. The thick, umber-tinted shinbone was from the leg of a steppe bison; the fragile-looking chocolate-brown jawbone with teeth once belonged to a diminutive Pleistocene horse. The two-foot-long, walnut-colored fragment of lower jawbone with a grinder still attached—that, I knew, could only have come from a mammoth. These exotic creatures once walked the ground that I was walking; perhaps they drank from this very spot at the river.

I switched with Zimov, dragging the boat and watching as he prowled along the beach. I couldn't wait to get back out there, and when I did, I started paying closer attention to the cliff face, where erosion from the summer thaw had liberated these bones from the *yedoma*. Occasionally there would be thin curtains of ice near the top of the bank. These were ice wedges, formed where the frozen ground had contracted and water had seeped in during the spring thaw. Erosion had exposed the ice wedges to the air. Mammoths would never be found in ice wedges; it would be the same as finding a fossil in an icicle. But once those icy curtains melted, frozen remains trapped for millennia might once again see daylight.

Bones and tree branches were sticking out of the riverbank, as if they had been victims of an ancient landslide. Scaling the cliff with my eyes, I suddenly noticed a frayed brown tangle. It looked like coconut husk—a crazy thing to think so far north of the Arctic Circle. I edged up to the bank, mindful that at any moment the unstable cliff face above could collapse and bury a *Homo sapiens* for future mammoth hunters to unearth. Quelling my bout of taphephobia, I tugged gently at the strands, hoping they were attached to a hulk that lay down for the last time in the Pleistocene Epoch. Snowflakes fluttered from the drab sky, a bleak warning of an early winter. In a few weeks the riverbank would be frozen solid. I could feel a discovery slipping away.

When I called to Zimov, he lumbered toward me, splashing through the shallows, mosquitoes caught in his thick beard. He stepped around a rust-colored mammoth femur and ripped a few strands of the ancient hair free. He rubbed it between his fingers and thumb. Only a neophyte, I trembled with anticipation. *Is it mammoth?*

"Steppe bison," he said.

For a moment I was crestfallen, then I recalled Zimov's earlier comment. I had to wait for the mammoth to find me.

— — —

The air had grown colder, so we had to endure a bone-chilling ride back to the station. After about fifteen minutes, Zimov slowed the boat, and we drifted toward a settlement of perhaps a dozen homes on the river's left bank. A pudgy man appeared on the bank, and Zimov hailed him. A few minutes later, the man was racing toward us in his motorboat and pulled up alongside. A fisherman, he and Zimov spent a few minutes negotiating a price for a 4-foot sturgeon lying frozen in the bottom of his boat. The fish would be tonight's dinner.

We had suffered another fifteen minutes of agony in the biting wind when we got lucky: we came upon a barge chugging toward Cherskii. Zimov knew the captain, who let him tie our boat to the side of the long, flat vessel. We passed the next few hours in comfort, sipping hot, sugary tea and munching on fish sandwiches in the heated cockpit.

Returning to Cherskii a few hours later, Zimov cleaned the sturgeon and cut its fatty flesh into chunks to marinate. Later we'd skewer the chunks and cook them over a fire.

I had some free time before our sturgeon *shashlik,* or shishkabob, to explore the grounds of the scientific station. Gray clouds hung low in the sky, and the air felt more like October than August. At the south end, Zimov and his son had built a wooden chapel with a green roof and a cupola on the edge of a bluff. The empty shelter could only hold a few people, but that was just as well. A far more powerful experience could be had outside the chapel, looking across the river at the vast larch-studded wetlands and snow-covered Mount Panteleikha, its upper half obscured by clouds. With my back to the station, there was no sign of humanity other than a derelict fishing vessel grounded on the near shore, its communications tower pitched toward the river. As I gazed ahead, the only sound I could hear was that of my lungs filling and emptying.

After dinner we drove into town to find Goto and the others. It was almost 11:00 P.M., and the sun was just dipping below the horizon. Zimov and I found Goto's ship at Cherskii's Green Point Harbor. The ship seemed abandoned, but descending belowdeck we found the Japanese sitting around the mess table drinking vodka in a haze of cigarette smoke. They were celebrating. "We succeeded!" Kobayashi whooped. "We found a mammoth!"

Well, not exactly. After a few toasts, Goto revealed that earlier in the day he had unearthed a chunk of mammoth skin. He and Kobayashi led us back up to the deck to show off their trophy. Goto grabbed a blue plastic tarp and, like a magician, yanked it away with a theatrical flair. Underneath was what looked like a rumpled gray bath towel. It was stiff, still frozen, but so much hair had fallen out that it might have been mistaken for elephant skin. It exuded the odor of the *yedoma*—the mangy smell that had assaulted my nose on the beach that day—only it was more potent. Lazarev, a short man with a graying mustache and intense black eyes, was certain the skin was from a mammoth's rump.

I would have been happy to have found a mammoth's rump that day, but I didn't see that the Japanese team had much cause for celebration. After two field seasons and hundreds of thousands of dollars, they had come away practically empty-handed: Goto had failed again in his hunt for sperm, while Iritani was leaving Siberia without the precious cells from a mammoth's reproductive organs. Iritani knew that the odds of this piece of skin containing cells suitable for cloning were vanishingly small. He insisted that the chance was not zero, but the forlorn look in his eyes suggested otherwise. My feeling was that this particular animal would not rise from the grave.

But more than a thousand miles to the west, on the Taimyr Peninsula, an Arctic tour guide–cum–mammoth hunter was gearing up for a project that offered the best hope yet of providing fresh-enough mammoth DNA to use in a cloning experiment. Most members of Bernard Buigues's team were not interested in bringing a mammoth back to life; they wanted to learn more about how the mammoth lived and why the species went extinct. But they were preparing to excavate a block of frozen tundra containing, according to radar measurements, what appeared to be an entire mammoth carcass. A mammoth may not have revealed itself to the Japanese team, but one had found Buigues.

THE RAT BENEATH THE ICE

The North Pole is a long way from anywhere. And it's an especially long way from where Bernard Buigues was born on the edge of the Sahara. His parents worked his grandfather's farm near Fez, Morocco, until 1962, when the seven-year-old and his family moved to Toulouse, in southern France. There, he recalls, people would look at the boy with blue eyes and blond hair and say, "Ah, yes, he must be from the north." The irony wasn't lost on the boy. "I would smile inside," Buigues recalls. "I *was* from the north—North Africa."

At university, Buigues started out studying medicine but in the third year, he says, he got cold feet. "I didn't want to be rooted to a community," he says about his decision to switch to philosophy before dropping out. Yet he admits he has a hard time leaving either of the two places he now calls home, Paris and Siberia.

After bouncing from one odd job to the other, Buigues's wanderlust and fascination with cold places eventually took him to Antarctica, where he served as the director of public relations for the Transantarctica expedition from 1989 to 1990. In this grueling adventure, his compatriot Jean-Louis Étienne, the first person to reach the North Pole alone without the help of dogs, spent more than seven months skiing from the tip of the Antarctic Peninsula to the South Pole, then on to Russia's coastal Mirny research station, on the other side of the continent. It was during that expedition that Buigues befriended a daring French mathematician, Christian de Marliave, who coordinated the logistics for Transantarctica. In October 1982, de Marliave and several friends set sail from France on a 12-meter steel-hulled boat they had built, making a two-month voyage to the Antarctic Peninsula, where he clambered all over the rugged terrain. Again feeling the pull of the icy continent, he returned five years later to conduct bird surveys for the British Antarctic Survey. After the Transantarctica expedition, when Buigues wanted to start a business that would take tourists to the Russian High Arctic and to the North Pole, he knew that de Marliave would be just the man to run logistics for him.

He settled on Khatanga, population 4,000, as his Siberian base of operations because of its proximity to the High Arctic destinations he advertised and because the local officials welcomed his business. Khatanga, Cherskii, and other Arctic settlements were little more than trading outposts before the military built up the country's perimeter defenses, placing radar installations and bases from Murmansk, near the border with Finland, to the village of Provideniya across the Bering Strait from Alaska, nine time zones away. The towns that provided the infrastructure—the hospitals, repair shops, and restaurants—flourished. During the Cold War, while most Russians had to pounce when consumer goods like shoes or light bulbs were suddenly and fleetingly available, Arctic dwellers enjoyed a steadier supply of goods and food and drew salaries at least three times higher.

Some European Russians, however, were seduced not by material goods but by the Arctic's open spaces. "It was a real opportunity for me to become free and travel," says Boris Lebedev, who studied in a military institute and did odd jobs before moving from Moscow to the Taimyr twenty-five years ago. His first job was as a hunter in a collective farm in a small Dolgan village 120 kilometers southwest of Khatanga. His life was catching foxes, fishing, and hunting. Lebedev, his head shaved and wearing a mustache, now lives in Khatanga and works as a ranger in the *zapovednik,* or nature reserve.

One dark side of this buildup was Joseph Stalin's campaign to corral the indigenous peoples who subsisted on hunting, fishing, and reindeer herding in the Arctic. Families accustomed to living in small villages and ranging hundreds of miles every year with their migrating reindeer were forced to settle in towns. There they often had little choice but to take the menial jobs rejected by the European Russian transplants. Like the fate of the American Indians in the nineteenth century, a thirst to subjugate drove this relocation. While Stalin never rationalized the cruelty he inflicted on his own citizens, he stood to gain from making the indigenous Siberians put down roots: by registering them as town residents, he could draft their young men into the military.

Such was the lot of the Dolgan people of the Taimyr, who also began to suffer economically, along with the European Russians, in the years leading up to the demise of the Soviet Union. By the mid-1980s, the country's sputtering economy made it harder and harder for the government to afford to fly in food or ship heating oil to the Arctic cities, most of which are inaccessible by road. And inflation over the last decade has only made things worse. During the Soviet days, says one

Khatanga resident, it took only two Arctic fox skins to barter for a round-trip ticket to Moscow. Now it takes forty.

With transportation and consumer goods throughout northern Siberia becoming scarcer and more expensive and salaries sometimes unpaid for months on end, the European Russians have been abandoning the Arctic in droves, leaving the decaying cities to the shrinking military presence and to the mining, shipping, and other industries still in operation. Khatanga has lost a third of its population in the last decade. Ignored by the central government, the indigenous peoples are returning to a subsistence lifestyle. According to the latest census by the administration of the Taimyr region—formally called the Dolgano-Nenetsk Autonomous District, after the two most numerous indigenous peoples living on the peninsula—exactly 272 Dolgans have gone back to nomadic reindeer herding. Their main purpose for coming to Khatanga is to barter for goods they can't themselves produce on the tundra.

In some ways, Buigues has become a surrogate for the Communist bureaucrats who used to fly out to small Dolgan villages with supplies. "He gives them sugar, tea, flour, every time he sees them. Bernard will buy shoes for kids but won't give money to men. That makes the women happy—the men would just buy vodka," says Vladimir Eisner, a meteorologist who came to Taimyr from western Russia twenty years ago. Eisner, who has worked as a translator for Buigues, now spends half his year living alone and hunting on the taiga and the other half in an apartment in town. But Buigues doesn't simply buy loyalty, Eisner insists. "When he smiles, it's genuine. He's a good person."

In autumn 1997, the Taimyr governor Nikolai Fokin told Buigues, out of the blue, about a nine-year-old Dolgan boy named Kostya who a few weeks earlier had spotted the tip of a tusk jutting from the permafrost north of Khatanga. Kostya's family had dug up that tusk and a second one they found below the surface and was preparing to sell them in Khatanga. Buigues was about to return home after wrapping up that season's North Pole expeditions. The official suggested that Buigues stay a few extra days and fly with him to the site where the tusks had been found.

Knowing almost nothing about mammoths, Buigues nevertheless visited the site, where he got a fragment of mammoth skull. That winter he considered the possibilities of mounting an excavation. Back in Khatanga the next February, he told the governor that he wanted to find the rest of the mammoth. "I was in a hurry. I was passionate about it," Buigues says. He flew by helicopter to meet the Dolgans on the tundra who laid claim to the site. Kostya's father, Gennady Jarkov, and Gennady's brother Gavril showed Buigues the specimens, each tusk more than 9 feet long and weighing around 100 pounds. Buigues bartered for the trophies. "It was not a question of money," he says. "It was a question of coffee, tea, and benzene" for their snowmobiles. Buigues began outlining an elaborate plan to disinter the rest of the mammoth.

The Jarkovs sold Buigues the tusks, but they were concerned about his intentions. Western and Dolgan cultures clash when it comes to the sanctity of the earth. Even digging a hole to plant a tree is forbidden, especially on the land outside town, which is considered sacred. "They think it's a sacrilege," says Eisner, who made the mistake one summer of planting several larch trees around his home 30 kilometers from Khatanga. "Dolgan kids came and broke the branches when they were brittle in winter, to make the trees die." Excavating a mammoth would be unthinkable. "We never disturb the bones in the earth. We only collect the bones from the beach," says Konstantin Uksusnikov, a Dolgan who hunts and fishes in Sobochnaya, a village dominated by a privatized collective farm with several thousand head of reindeer.

That taboo against unearthing mammoth bones has outlasted Soviet attempts to suppress cultural beliefs by educating the native children with the children of European Russians. While the plan succeeded in forging many good Communists, the traditions about the lost creatures survived.

Particularly reverent were the Yukagir, who live on a patchwork of thin forests laced with sloughs, lakes, and tussocky marsh between the Kolyma and Indigirka rivers. Guarding each of their shaman, liaisons with the spiritual world, is the spirit of a bear, a reindeer, or—if a shaman were highly influential—a mammoth, which they call the *xolhut*. No force is more powerful than the *xolhut aibi,* the mammoth shadow.

In contrast to many other indigenous peoples, the Yukagir in the nineteenth century thought the mammoth neither subterranean nor alive. They believed that the beast dwelled aboveground a long time ago, then went extinct. Their tradition held that "the creation of the mammoth was a blunder of the Superior Being," according to Waldemar Jochelson, an anthropologist who led an expedition to the region in the late 1800s. The *xolhut's* undoing, the Yukagir were convinced, was its size and strength. They believed the ravenous beast had stripped the land of trees, converting vast Siberian tracts to tundra. The denuded marshes and loose sand became a death trap, swallowing the heavy *xolhut,* then freezing it solid in winter. That explained why the tundra was the permanent crypt of countless *xolhut.*

If explorers were to raise the Ice Age dead, the consequences, from the Siberians' perspective, could be devastating.

Witness the death of Herz and the others involved with the Berezovka mammoth soon after its excavation. They paid a price for unearthing the king of the rats. On other occasions, explorers seeking cooperation from the indigenous Siberians were rebuffed for more prosaic reasons: they just wanted to be left alone by the often-imperious Westerners. The Baltic German explorer Gerhard Gustav Ludwig von Maydell, who collected three mediocre mammoth carcasses in three years of searching in northeastern Siberia, often complained about the lack of cooperation he received. In 1869, a year before his last find, he wrote that the Yakuts "whenever possible ... conceal all finds, fearing to be forced to work and provide haulage." Not even a bounty of up to a thousand rubles for solid leads would change their minds. "They thought the premium could not recompense them for all the troubles connected with the arrival of an expedition and with the travel of government officials," wrote the early-twentieth-century mammoth hunter I. P. Tolmachoff.

While unearthing mammoth remains continued to be taboo, the tusks lying aboveground could be gathered without fear of retribution. For more than two thousand years the ivory, called *mamontova kost* in Russian, had been shipped from Siberia south into China, Mongolia, and beyond via the Silk Road, a caravan route. Mammoth ivory was so valued that in the thirteenth century, the Mongol Khan Kuyuk reigned on an ivory throne. Only after the Cossack chieftain and former river pirate Yermak Timofeyevich conquered Siberia on behalf of the Russian Empire in 1582 did the ivory begin to make its way west to Moscow, then on to Europe. (The first documented mammoth ivory to reach western Europe arrived in London in 1616.) Most of the ivory that reached the market was collected by the indigenous people, who would find tusks sticking out of the tundra and mark their claim, then return to the site in warmer weather to see if their prize had melted free. They would then sell or barter the tusks in the nearest village.

Although the ivory trade was opportunistic and relied on warm summers for its boom years, industry records hint at legions of disinterred mammoths. During the first half of the nineteenth century, as many as 32 tons of ivory were sold each year in Yakutsk, the wild frontier town that remains the hub of Russia's diminished ivory trade. Trading picked up in the second half of the century, with more than 2,700 tusks exported to England alone in a two-year period in the 1870s. Fine mammoth ivory used for carving scrimshaw fetched up to $10 a pound in Europe at the time. By the early twentieth century, experts say, the tusks of as many as 50,000 mammoths had been carried off to market. Siberia was so rich in ivory that the *Encyclopaedia Britannica* of 1911 called the land "inexhaustible as a coalfield and in future, perhaps, the only source of animal ivory."

Hoping to leverage the ivory trade to suit their own interests, mammoth hunters with the Russian Imperial Academy of Sciences sent a notice in 1860 to the regional Siberian governments: they would pay a bonus to anyone who informed them of the location of an intact skeleton of a mammoth or other gigantic antediluvian animal. While mammoth hunters in Siberia had found portions of nearly two dozen carcasses in a half-century of searching, none by that time was as valuable to science as the Adams mammoth. The scientists offered 100 rubles for a tip, an additional 50 if the find proved important enough to excavate, and 300 rubles if any flesh or skin was still attached to the skeleton. "Do not report any groundless gossip," the notice cautioned.

It rested on local officials, most with no scientific training, to screen any claims before forwarding them to the academy.

Sometimes that spelled trouble. In 1877 the academy spent a hefty sum—1,000 rubles (at that time enough to buy a herd of cows)—for a team to check out a report of a mammoth carcass in the Kuznetzki Alatau Mountains, in southern Siberia. Lending credence to the report, the villager who found the carcass was said to have eaten a piece of its skin. But when the scientists arrived, they found that the skin was no more than a brownish patch of mountain leather, a type of asbestos in which the mineral forms thick sheets. (The chastised villager claimed it was possible to eat anything if it was dipped in butter.) On many occasions, wrote Tolmachoff, academy scientists, "after long and hard travel over thousands of miles, arrived at the places only to dig out a few bones and poor remnants of soft parts."

— — —

Bernard Buigues did not want to repeat that mistake. In April 1998, accompanied by the filmmaker Jean-Charles Deniau and a photographer as well as his local Cerpolex staff, Buigues began excavating at the site the Jarkov brothers had found, near the Bolshaya Balakhnya River. After three days of shoveling snow and chipping at frozen dirt, Buigues found the damaged crown of the mammoth's skull and the fractured sockets from which Gennady Jarkov had pulled out the tusks. In the months since, the skull muscles and ligaments had been devoured by Arctic fox or other scavengers and further cleaned by bacteria. "Many times, I thought, What am I doing here, spending this money?" Buigues says. "I was really depressed." Then he had an idea: he would connect his hair dryer to an electric generator and melt the sediments containing the mammoth. After only several minutes of blowing, he found a swatch of hair still attached to skin and, beneath that, meat. Buigues knew he was on to something big.

How big he didn't know. To get a better idea of how much of the mammoth might lie underground, Buigues flew in a Swedish expert who brought ground-penetrating radar. This device, which sends radio waves into the ground and charts the reflections that bounce back, revealed a large anomaly: an oddly shaped mass, about as large as an adult mammoth, that was less dense than the surrounding soil and ice. While far from certain, the readings indicated flesh and bone.

That was enough for Buigues to reach out for more scientific muscle. Stocky, blond Dirk Jan "Dick" Mol, a highly respected Dutch amateur paleontologist, was organizing an international mammoth convocation in Rotterdam in August 1998. He remembers well his phone call from Buigues one evening. "I had never heard of him before; I didn't know anything about what was going on in Siberia," Mol says. Buigues described his preliminary excavation and asked if he could visit Mol in Amsterdam—in two days. Mol agreed, and in preparation searched the scientific literature on mammoths for Buigues's name. "I never saw a reference to Bernard Buigues. I was wondering, what kind of person is this?" Mol's unease grew when he met Buigues at the airport. "I started speaking about mammoths, and noticed that this man didn't know anything about them at all!"

At Mol's home, Buigues showed Mol pictures from the site, including several depicting the upper and lower jawbones still hinged together. This suggested that after the mammoth died, its carcass had been buried quickly by silt. It didn't appear to have lain around to be picked apart by scavengers, which would have scattered its bones. "I became very enthusiastic," Mol says. He also realized that Buigues had done a careful job with almost no technical support. "I told Buigues, 'It's very wonderful what you are doing.'" Buigues then revealed that he was doing a film for French TV, and he offered to take Mol to Siberia to consult on the project. The next month Mol went to Paris, where he and Buigues held a press conference to announce their project and show a few minutes of the film. There, Buigues gave Mol a fragment of skull and a dime-size piece of skin, which he later took, along with a few strands of hair, to the University of Utrecht for radiocarbon dating.

Radiocarbon measurements can date samples precisely. The moment an organism dies and stops inhaling carbon dioxide, the abundance of carbon isotopes in its tissues, fairly consistent across species, begins to shift as minuscule amounts of radioactive carbon-14 decay to stable nitrogen. Thus carbon-14 is a postmortem clock that stops ticking only after all the radioactive atoms are gone. With half of the carbon-14 atoms decaying every 5,730 years, enough carbon-14 remains in a sample to accurately measure age back at least 50,000 years. Beyond that, the scarcity of remaining carbon-14 yields an ambiguous date.

Buigues and Mol both believed that the rest of the mammoth was still in the permafrost. Buigues had considered a traditional excavation, which would blast the permafrost with hot water in summertime. "But that would have destroyed the scientific information," says Mol, who with Buigues decided to undertake a riskier, and far more expensive, endeavor: chisel out the chunk of permafrost containing the remains and airlift it to a cold room in Khatanga for study. "It would be the first time that scientists would have access to materials in which the frozen chain was not broken," Mol says. He wasn't preserving mammoth DNA for a cloning experiment; he had no desire to see the mammoth brought back to life. Mol objected on purely philosophical grounds. Because the habitats that once supported the mammoth had disappeared, he thought it would be cruel to resurrect a species that could live only in a zoo. But Mol did agree to share tissue samples with cloning researchers, which delighted Kazufumi Goto and his team. Fresh from their failed expedition, they were eager to see what Buigues and Mol had found.

~ ~ ~

Mol's fascination with fossils had been kindled when a grade school geography teacher in his hometown of Winterswijk showed him a collection of sea urchin fossils dating from tens of millions of years ago. As a teenager, he accompanied his uncle to the Leiden Museum, where curators showed him backroom collections of mammoth bones gathered from the North Sea. "From that moment on," he says, "I decided to collect the remains of Ice Age mammals."

The eighth of nine children whose father was a police officer, Mol could not afford education beyond high school. But his decision to join the customs service in 1974 proved a major boon. That same year, the Netherlands implemented the Convention on International Trade in Endangered Species of Wild Fauna and Flora, which gave customs officers sweeping interdiction powers. Mol, who trained as a CITES treaty specialist at Amsterdam's Schiphol Airport, spends so much time studying bones and interacting with professionals in his work that he doesn't regret missing out on an academic career. Nor has he ever found a colleague from academia reluctant to collaborate with him. "Many scientists don't even know that I'm a customs officer," he says.

His employer has also benefited from his passion. A traveler from Hong Kong once declared that an ornate sculpture of a Chinese god was carved from woolly mammoth ivory. (Because mammoths are extinct, their parts are not protected under CITES, unlike those of living elephants.) "I told him, 'I do not believe you, and I'll tell you why,'" Mol recalls. The object was clearly carved from a straight tusk without any cracks, he says—a dead giveaway, since mammoth tusks are curved and even the best-preserved tusks have cracks. Mol gave the man a choice: either ask a court to order a scientific test to determine the ivory's age or pay a steep fine. The would-be smuggler paid up.

Mol started building a collection of Pleistocene fossils twenty-five years ago, when he met a few fishermen who agreed to bring ashore whatever they pulled from the sea. Since then he has amassed some 15,000 specimens, turning his suburban townhouse into an Ice Age museum. Sculptures of mammoths, several of plastic and one of pewter, sit on some bookshelves, while waist-high mammoth femurs lean against a wall. In one living room corner is a mammoth's lower jawbone that Mol would have trouble getting his arms around. His most beautiful trophies include exquisitely preserved mammoth tusks, teeth, and a Pleistocene musk ox horn sheath. The rest, cleaned, labeled, and numbered, fill some 600 styrofoam boxes the size of footlockers. When Mol's daughter moved to The Hague a few years ago, her room was soon taken over by bones. "I'm waiting for my son to leave the house as well," he says.

Mol's exploits recently brought him an unexpected honor. In April 2000, when he was laying the groundwork in Khatanga with Bernard Buigues for that summer's expedition, he received a phone call from the Dutch ambassador to Russia, saying that he would receive a knighthood from Queen Beatrix. He was being honored for his highly visible role in educating the public about paleontology, both as a spokesperson for the Buigues expedition and for Her Majesty's Customs Service. The next month, when Mr. Mammoth became Sir

Mammoth at a quiet ceremony at the Ministry of Finance, the honor brought practical results. "I asked my boss if it would be possible to get some extra holiday time to spend in Siberia with the expedition," says Mol, who had already used up his generous allotment of three months' annual paid leave. "He said, 'Of course. How much time do you need?'"

No other country, perhaps, embraces amateur paleontology as warmly as the Netherlands, and Mol's success has lent credibility to this thriving community. "Vertebrate paleontology as an academic subject now hardly exists here," says Jelle Reumer, who notes that hot fields like genetics tend to get the few new academic positions created at universities in his country. "Amateurs help fill that gap." Amateurs and professionals meet four times a year to swap stories, and together they produce a journal, *Cranium,* featuring the results of their collaborations. Mol may be the most accomplished of the Dutch amateurs, says John De Vos, a curator at the National Museum of Natural History in Leiden. "He knows every mammoth specimen in Europe. He's crazy! He's obsessed!"

Mol has built his collection largely on donations from kindred spirits: a network of fishermen, also bone collectors, whom he visits on occasion to trade fossils and stories. Few trips onto the North Sea fail to dredge up the past: antique glass buoys the size of beach balls, for example, or Meiji soy sauce bottles, reminders of the thriving early Dutch trade with Japan. One time a net brought to the surface a seventeenth-century bronze cannon.

The biggest threat to Mol's hobby is the mushrooming fossil trade in the Netherlands. In the late 1800s, most fishermen saw Ice Age bones as a nuisance, to be shoveled overboard. "Most fishermen are not amateur paleontologists, they are mostly interested in the money they get for the fossils," Mol says. So Dutch and Belgian dealers began offering a financial incentive to the fishing vessels. Some scarce, well-preserved bones are hot commodities: on the open market, molars from the southern mammoth, an early representative of the mammoth lineage from which only about sixty molars have ever been recovered from one section of the North Sea —the *Oosterschelde,* or Eastern Scheldt, off the Dutch province of Zeeland—fetch up to $500. Less scrupulous dealers, Mol asserts, sell scientifically valuable specimens to the highest bidder instead of following the unwritten rule of first offering such finds to museums or research institutes. Mol suspects that some specimens are disappearing into private collections before experts can give them even a cursory glance. "This is very bad for science," he says.

As his acumen grew, Mol started to see things in the sea fossils that the professionals had missed. His biggest coup was being the first person to spot a flaw in a well-publicized 1993 report in the journal *Nature* claiming that the freshest mammoth remains ever found belonged to dwarf males. After seeing the bones of these individuals, which had died only 3,700 years ago on Wrangel Island in the East Siberian Sea, Mol realized they belonged not to dwarfs but to old female mammoths, which, like elephants, were smaller than the males. "Most scientists want to discover something new, something spectacular," says Mol. "But once you have this in the literature, it's very hard to dispel." He did his best to dispel the dwarf mammoth idea, however, in a report delivered in 1999 at the Second International Mammoth Conference in Rotterdam. As a result, California's Channel Islands won bragging rights to the only true dwarf mammoths known from the fossil record.

Mol also figured out that there are three mammoth species represented among the North Sea fossils, illustrating that the sediments encompass millions more years than previously thought. After persuading the fishermen to chart the original location of each fossil, Mol was able to map the areas of different groups of species. "We now know exactly where we can find which remains," he says.

The North Sea bones have also helped Mol and his colleagues paint a fuller picture of the evolution of mammoths in Europe and Asia. We now know that the mammoth originated in North Africa around 4 million years ago. Bones of this species, called the African mammoth (*Mammuthus africanavus*), have been found in Algeria, Chad, Morocco, and Tunisia. The African mammoth migrated north, either across the Strait of Gibraltar—where a narrow strip of land once connected Africa and Europe—or through the Sinai Desert and the Middle East. By the time this beast died out at the start of the Pleistocene Epoch, around 1.8 million years ago, it had given rise to a new species, the southern mammoth (*Mammuthus meridionalis*). Notwithstanding its name, the southern mammoth, which grew to 14 feet at the shoulder, never reached the Southern Hemisphere but was adapted to life on the savanna—a warm, dry grassland with few trees. Based on its low-crowned teeth with fewer enamel plates than those of later mammoths, it's reasonable to imagine that southern mammoths had a diet similar to that of their distant cousins, the mastodonts, woodland creatures that preferred twigs and leaves to grass.

Bones from these older mammoths are found in certain sections of the North Sea's bottom, primarily in the north-south-running Deep Water Channel, where the sediments date from the end of the Pliocene into the early and middle Pleistocene. Radiocarbon dating is useless on these bones, and techniques to measure time by analyzing crystals in the surrounding sediment are likewise pointless, because no one knows which sediments, exactly, the dredged bones came from.

The best clues to their age come from comparing the assemblage of animals found in a particular patch of the North Sea with digs on land, where sediments can be dated and tied to specific fossils. For example, the muck at the bottom of the Eastern Scheldt dates from 1.6 million to 1.8 million years ago. During that period, called the Tiglian, today's estuary and other parts of the southern bight of the North Sea between England and the Netherlands were dry lands inhabited by southern mammoths, squat mastodonts, giant deer, and the saber-toothed cats that preyed on them. A site in Chilhac, France, with fossils of both a mastodont called *Anancus*, whose straight tusks up to 13 feet long extended nearly the entire length of its rhino-like body, and the southern mammoth, dates to around 1.9 million years ago. Since bones of these two species aren't found together anywhere else in the world except in the Eastern Scheldt, Mol and his colleagues think the two sites date from the same time.

At the beginning of the Pleistocene, many species went extinct, including *Anancus* and a three-toed horse (modern horses have a single toe, the hoof). The extinctions made room for other species, like the southern mammoth, which for the next 2 million years fanned out across Europe and northern Asia, eventually lumbering across the Bering Land Bridge, a now-submerged portion of Beringia—half the width of the United States—that connects Asia and North America. With each glacial advance, the Northern Hemisphere, on the whole, became colder and colder and drier and drier, Mol says. During the middle Pleistocene, from around 700,000 years ago until 125,000 years ago, lush savannas were supplanted by harsher steppes. Eventually, Europe became "one big, vast steppe," Mol says. The southern mammoth waned, and in its place arose the steppe mammoth (*Mammuthus trogontherii*), a creature better adapted to life in cooler climes.

While the teeth of southern mammoths formed from about a dozen plates pressed together, the steppe mammoth had more plates and a higher crown, enabling it to grind the steppe's tougher grasses better.

Around 300,000 years ago, somewhere in Eurasia, the steppe mammoth spun off the woolly mammoth (*Mammuthus primigenius*), which marched west as far as Spain and, following the southern mammoth, east across the Bering Land Bridge into the New World. The woolly's teeth adapted to the extreme, adding more and more plates for grinding tougher grasses. It had twice as many plates in its molars as the southern mammoth, it was short—averaging less than 9 feet at the shoulder—and it grew its hallmark downy undercoat and long bristly hairs. The seabed southwest of the North Sea's Brown Bank is rich in bones of woolly mammoths and other creatures from the late Pleistocene.

Having filled out the Eurasian branches of the mammoth's family tree, Mol and other scientists have increasingly turned their attention to solving the mystery of the creature's demise. Once one of the most abundant mammals on the steppes of Eurasia and North America, woolly mammoths suddenly died out at the end of the Great Ice Age, whereas contemporaries like the reindeer and the musk ox survived. Scientists could debate endlessly over which of the mammoth's enemies—climate change, hunters, or disease—conquered the species. The only way to settle the question is to unearth vital clues.

Because the Taimyr mammoth that the Jarkovs found lived during the species' heyday in Siberia, it alone could not solve the riddle. Indeed, the animal had no bearing on one hypothesis: that human hunters wiped out the mammoth. Support for this notion would have to come from finding troves of mammoth bones with telltale signs of butchering—abundant in North America, but harder to find in Siberia.

But if the Taimyr mammoth were nearly intact, it could be used as a yardstick for assessing evidence from mammoths that died more recently, when their populations were plummeting around 10,000 years ago. The mammoth enticed those scientists who speculated that an apocalyptic disease annihilated the species, for the mammoth's lungs, blood, or other frozen tissues might hold traces of an infection—perhaps a relatively benign version of a pathogen that later evolved to wipe out the species. And Mol knew that the animal's stomach should contain half-digested plants that would help scientists reconstruct the environment on the Taimyr Peninsula 20,000 years ago. If more recent specimens had less nutritious plants in their stomachs and appeared undernourished, that would support the notion that a climate-driven shift in vegetation killed the mammoth—a hypothesis that the Dutch knight and many others clearly advocated. Oddly enough, it was a knight on the other side of the North Sea a century earlier who was the first to explore this idea systematically, turning his vision of a mammoth-blighting climatic upheaval into a personal crusade.

A DEADLY CHILL

Soon after Georges Cuvier showed that the mammoth had gone extinct, people started wondering why. Why had the mammoth been banished from the world forever? What blow could have felled this mighty creature? In the nineteenth century, the first of three hypotheses emerged: that mammoths were the victims of a sudden climate change that made life unbearable and ultimately impossible to sustain.

By the mid-1800s, most geologists had come to embrace uniformitarianism, which held that the laws of nature worked no differently in the past nor would they work differently in the future. The father of this theory, the English geologist Sir Charles Lyell, explained it as follows: "The forces now operating upon the Earth are the same in kind and degree as those which, in the remotest times, produced geological changes." Shaping the landscape were eternal processes: cliffs eroding under a pounding surf, fallen leaves decaying and forming soil, earthquakes tearing the land asunder, volcanoes spewing lava that hardened into new land. The laws of nature were immutable.

However, Sir Henry H. Howorth, an antiquities scholar, thought that nature had the potential to behave more violently than people realized. Look at the moon, he challenged, and explain how the craters got there. There was simply nothing happening on the placid satellite that could account for its scoured face. Attributing crater formation to the relatively benign geological processes of modern times, argued Howorth, was like attributing the pitted cheek of a grown man "to the operations of his ordinary life," when his scars in fact resulted from a childhood case of chicken pox. Whatever agent had battered the moon was as elusive as a childhood infection of yesteryear.

Howorth beat up on uniformitarianism in preparation for invoking an ancient cataclysm as the explanation for the demise of the mammoths. In a series of articles in *Geological Magazine* in the early 1880s, Howorth argued that the floodwaters from torrential rains had swept up the mammoths and their contemporaries such as woolly rhinos, snuffing out scores of species at once. When the waters receded, the waterlogged carcasses lodged in crevices in the tundra, and global temperatures plummeted in a matter of days, freezing the beasts where they lay. As evidence, he quoted the writings of Siberian explorers, whose investigations of mammoth and rhino remains often indicated that the beasts drowned or suffocated.

Howorth recounted one explorer's speculation about the origins of a thick layer of sand and loam near Yakutsk. It was "deposited from waters which at one time, and it may be presumed suddenly, overflowed the whole country as far as the Polar Sea." Through his provocative writings, Howorth resuscitated the explanation for the mammoth's extinction proffered by the seventeenth-century Russian settlers in Siberia. That cause, he wrote, could be none other than the Great Flood described in the Bible.

This idea, Howorth knew, would outrage many of his peers. Here was a scholar putting stock in Noah's deluge at a time that scientists were waging a bitter war with theologians over the validity of Charles Darwin's theory of evolution. Howorth girded for battle. "I expect and shall welcome the keenest criticism," he wrote in an 1887 treatise, "The Mammoth and the Flood." He urged readers to be swayed neither by the tyranny of scientific orthodoxy nor the dogmatism of theology. While it is rational "to apply to the Bible the same canons of criticism and analysis we would apply to any other book," he wrote, it is "irrational ... to refuse credence to a story *because it is contained in the Bible*" [original emphasis]. Howorth, it appears, did not believe in a literal interpretation of Noah's flood, but he did feel that the story was rooted in reality because many cultures, including the ancient Egyptians, had similar flood stories. "These traditions generally agree in placing a great catastrophe, involving widespread destruction to animal life, at the verge of human memory," he wrote.

Ironically, even Lyell, a staunch defender of uniformitarianism, had proposed a similar watery death to explain the frozen mammoth carcasses in northern Siberia. The remains in the High Arctic, he reasoned, were animals that lived in a better habitat in central Asia that were swept into the Arctic by Siberia's north-flowing rivers. Lyell and many others were convinced that in Siberia, the modern climate mirrored the prehistoric—and that this climate was inhospitable to mammoths. It was so cold in the High Arctic, noted the nineteenth-century Russian explorer Baron Ferdinand Petrovich von Wrangel, that the wild reindeer "withdraws to the deepest thicket of the forest, and stands there motionless as if deprived of life." How could a mammoth, which had to eat constantly to maintain its weight, survive in such a hostile land?

After Howorth's treatise appeared, a handful of scholars argued that he had made a strong case, and they invoked bizarre geological phenomena to account for a global flood. "Let us imagine that the Earth, once intensely heated, had slowly cooled down and shrunk," wrote Brownlow Maitland in the *Quarterly Review* in 1888. As the earth's interior condensed, the crust would not have shrunk, Maitland proposed, leaving a gap. Finally the crust would come crashing down, "starting some mighty oceanic wave to roll with desolating fury over neighboring lands."

Most scientists, however, derided the biblical revival. While applauding the iconoclastic Howorth for compiling a valuable encyclopedia of mammoth lore, one geologist complained that "the conclusions he draws seem quite unwarranted by the facts, and are thoroughly opposed to the view of most modern geologists, that vast changes need a great lapse of time." That critic among others didn't think it plausible for temperatures across the Northern Hemisphere to have dropped precipitously in a matter of days, thereby discrediting Howorth's hypothesis of a "quick freeze" encasing the drowned mammoths. One geology professor, W. Boyd Dawkins, accused Howorth of disposing of evidence contrary to his views, saying: "We are not in a court of law, but in a court of science, where the wig and the bands of the special pleader appear to me to be out of place." With few of his peers willing to accept that the flood was more than an allegory, Howorth by some accounts became a tragic figure: "a fervent, although solitary, advocate" of the hypothesis until his death in 1923, according to the Russian expert I. P. Tolmachoff. One independent scholar, Hans Krause of Stuttgart, Germany, has resurrected and extended Howorth's ideas, but he has won few converts in the scientific community.

Howorth gained a posthumous measure of redemption in 1993 when two Columbia University scientists uncovered evidence that a titanic prehistoric flood really did occur. About 7,000 years ago, water breached a narrow strip of land separating the salty Mediterranean Sea from the fresh Black Sea. With the thunderous sound of a hundred Niagara Falls, the water crashed through what is now the Straits of Bosporus, flooding low coastal areas around the Black Sea at a rate of nearly a mile a day. That cataclysm, revealed by evidence such as radar echoes of the now-brackish Black Sea's ancient coastline, may have given rise to the legends of a deluge, including the story of Noah's Flood.

But Howorth was only partially vindicated: the Black Sea flood had nothing to do with the demise of the mammoths. Its putative occurrence came at least 2,000 years after the last steppe mammoths east of the Black Sea had died out. And it would have inundated only the lowlands of modern Ukraine and other countries bordering the sea.

Howorth's notions that nature was not constant and that past climates could differ greatly from current ones got a boost when fossils of hippos were discovered in England in the 1800s. The obvious question, according to Paul Martin, one expert on Pleistocene extinctions, was: "How can you possibly have a tropical animal like a hippo up in Trafalgar Square?" Perhaps the winters were warmer, scientists concluded.

But perhaps they were colder, too. And if the climate were changing abruptly, would mammoths—and other animals—

have been woefully unprepared? At the forefront of this revisionism was Henri Neuville, a biologist who in 1910 started an investigation of mammoth physiology at the Paris Museum's laboratory of comparative anatomy. Neuville had not set out to tackle the question of why the mammoth went extinct until an extraordinary opportunity presented itself.

In 1907 Konstantin Vollosovich, a Russian geologist, was alerted to the location of a carcass of a small male mammoth on Bolshoi Lyakhovsky, one of the New Siberian Islands off Russia's northern coast. Vollosovich borrowed heavily to pay for an excavation that proceeded slowly, due to a shortage of sled dogs to haul the remains some 1,500 miles south, to Yakutsk. The mammoth—a nearly complete skeleton with skin, an ear, an eye, and mounds of hair—was intriguing because of its coat's patchwork of colors, from straw yellow to roan to dark brown. Vollosovich speculated that when it died, the mammoth was molting, losing its summer coat and growing its winter pelage. By 1910 the mammoth's remains were safely stowed in a refrigerator in St. Petersburg, but Vollosovich was still unable to persuade the Imperial Academy of Sciences to reimburse him for the expedition. A friend, Count Ivan Stenbock-Fermor, agreed to bail him out. To his surprise, however, Stenbock-Fermor turned around and donated the mammoth to the Jardin des Plants in Paris. Tolmachoff, who spoke with Vollosovich before his death, learned why: "The reason for such a generous gift was the hope of being decorated with the Légion d'Honneur and, in the capacity of a possessor of this decoration, of having at his funeral a military band playing."

From this unexpected gift—the only virtually complete mammoth skeleton from Siberia ever allowed to leave Russia permanently—the Paris Museum obtained some of the well-preserved skin, which Neuville began to examine. In the process, he corrected one common misconception: that mammoths over many generations grew a thicker hide to withstand the cold. According to his measurements, mammoths and elephants were equally thick-skinned.

But the mammoth's skin, Neuville argued, was also one of its greatest handicaps. Examining it under a microscope, he concluded that mammoths, like elephants, lacked fat-secreting sebaceous glands. "Everyone knows to what degree the presence of the grease produced by the sebaceous glands renders wool resistant and isolating," Neuville wrote. Only a few species of mammals, he pointed out, lack fat glands, and they live in the tropics; besides elephants, he knew of only the two-toed sloths of Central and South America and the golden moles of Africa. Since sloths are acutely sensitive to cold and damp, Neuville concluded that the mammoth wool, lacking sebum, "furnished only a precarious protection against cold." Seven decades later Soviet scientists would prove Neuville wrong by demonstrating that mammoths indeed had numerous fat glands. "[That] discovery meant that mammoth wool and skin had normal waterproof qualities," observed the Ukrainian zoologist Pavel Putshkov, a Pleistocene specialist.

The mammoth's tusks, however, gave Neuville a better argument for poor adaptation to cold. He viewed the adult male mammoth's tusks, which curved inward and back toward the body, as deformed versions of straighter, shorter elephant tusks. Male elephants are territorial and often belligerent toward other males, brandishing their tusks as weapons in fights over females. To keep their tusks straight and sharp, elephants rub these overgrown incisors against trees. But in northern Siberia during the Pleistocene there were few trees to serve as tusk sharpeners, Neuville asserted, so the tusks grew wild. "Rather than efficient weapons," he wrote, "they appear to have been only encumbering accessories."

Whether a mammoth's tusks were in fact adaptive is still debated. African elephants, for example, use theirs to uproot small trees to clear paths and scrape cave walls to liberate salt and other essential minerals; males, in particular, wield tusks as weapons when dueling over mates and fending off predators. But the pronounced curve of male mammoth tusks rendered them useless for piercing and fighting. About the only thing that adult males were able to do with their tusks, scientists speculate, was sweep away snow to get at the grasses below. Young mammoths, however, might have been able to use their tusks—before they curled in on themselves—to break up ice to get at water to drink in the wintertime, according to analyses of fractured and worn juvenile tusks.

In contrast to the early-twentieth-century paintings of woolly mammoths striding vigorously across snow-whipped tundra, Neuville portrayed the beasts as degenerated elephants struggling for survival in an increasingly hostile realm. "The snow, the icy rains, could have penetrated the curious fur with which the animal was covered; the fur must then have transformed itself into a veritable cloak of ice, not merely in a superficial manner, but down to direct contact with the epidermis." The only characteristic that he deemed protective against the cold was the mammoth's tiny ears, which in shape resembled our own. (The largest mammoth ear found is just over a foot long, whereas African elephant ears measure 6 feet or more.) While acknowledging that his conclusions were based on limited observations, Neuville nonetheless reasoned that the mammoth's extinction was a case of a pitiable creature succumbing to inclement weather.

— — —

Neuville was close to the truth, according to those experts who today argue that the sudden climate change at the end of the Great Ice Age killed the mammoths. One reason they started to believe that global phenomena were at play was that many more species than mammoths were lost at the end of the Pleistocene. Describing the devastation in terms of animals weighing more than 20 pounds, on average, Eurasia lost 28 species; North America was hit even harder, losing 48. On both landmasses, three of every four species of megafauna —animals weighing more than 100 pounds—were obliterated. Besides mammoths, the victims included mastodons (as American mastodonts are called), saber-toothed cats, giant ground sloths, and Pleistocene horses. Then the destruction abruptly stopped. For the next 10,000 years—a much more stable period, climatically—not a single mammalian species disappeared from the North American and South American mainland. In Eurasia, only a single species—the musk ox—disappeared.

As rapidly as species were getting snuffed out at the end of the Great Ice Age, the dieoff was nowhere near as devastating as any of the five mass extinctions occurring in the 600 million years leading up to the Pleistocene. As one expert put it, the end of the Pleistocene was not a mass extinction but rather an extinction of the massive. Still, a closer look at the forces that operated on grander scales during the "true" mass extinctions, when untold numbers of species perished, might shed light on whatever calamity befell the mammoths and other Ice Age animals.

The two most recent mass extinctions involved large land animals and can serve as comparisons. The most devastating event took place at the end of the Permian Era, 251 million years ago, when 90 percent of the earth's marine species died out, paving the way for the rise of dinosaurs and mammals. Before the cataclysm, the earth formed a single supercontinent, Pangea. One provocative recent hypothesis ties this extinction to a titanic upwelling of molten rock from the mantle that fissured Pangea, setting the continental plates on a path toward the positions of today's seven continents. According to one hypothesis, the magma exploding through the crust released enough gases and dust to blot out the sun for years. Deprived of sunlight by the prehistoric smog, plants could not photosynthesize and died. The loss of plant life meant that little oxygen was being pumped into the atmosphere, allowing carbon dioxide and other greenhouse gases to accumulate and raise global temperatures. It was millions of years before the volcanic activity subsided enough to allow a few hardy species to give rise to diverse classes of life.

Another huge deathblow came 186 million years later, when the earth lost about a third of its species, from ocean plankton to the dinosaurs. Like the Permian extinction, the most likely culprits again were sudden temperature changes and curtailed photosynthesis. There's persuasive evidence for an extraterrestrial trigger: a massive meteorite crashed into the waters off Mexico's Yucatán Peninsula with a force that could have shot enough dust into the atmosphere to block the sun for months, strangling life.

The common theme underlying both extinctions is a dramatic shift in global climate that crippled plant life. Like a demolition crew leveling a building by dynamiting its supporting pillars, the climate changes struck at the base of the food chain, killing plants. Deprived of food and their main source of oxygen, herbivores perished, as did the carnivores that preyed on them.

The Pleistocene Epoch (the first of two chapters in the present Quaternary Period) did not go out with a bang caused by global volcanic activity or a gigantic meteor strike. But the era witnessed severe and sudden climate shifts, albeit more muted than the catastrophes that led to the mass extinctions.

Over the 2 million years spanning the Pleistocene, global temperatures swung from icehouse to hothouse at least twenty-three times. When temperatures fell into a prolonged slump, the polar ice caps crawled toward the equator, reached a maximum, then retreated. The cold snaps lasted anywhere from 40,000 to 100,000 years at a stretch, separated by relatively warm interglacial periods. Almost all mammalian species, including the megafauna, survived these climatic vicissitudes. To some experts, that's rather odd. "It seems impossible," notes Paul Martin, an ecologist at the University of Arizona, "that by the end of a long period of swings from cold to warm, and back again, we could run into just one more that would make all the difference in the world."

But something did go terribly wrong at the end of the last deep freeze, which began about 100,000 years ago. At the height of the Great Ice Age, known as the Wisconsin Glaciation in North America, glaciers covered much of the continent, extending into today's Montana, Kansas, and Pennsylvania, as well as parts of northern Europe, including Scandinavia. Besides being cooler, the air was apparently much drier, which nurtured grasslands across North America, Europe, and northern Asia.

Eventually, temperatures eased higher, and the glaciers began a slow retreat that lasted until the Pleistocene ended, around 10,000 years ago. Then the world really began to heat up: global temperatures, according to some estimates, rose roughly 5 degrees Celsius in as little as two decades. (Compare that to the rise of 0.75 degrees Celsius over the past century, attributed to global warming.) This transition from the Pleistocene to the Holocene Epoch can be considered a much milder version of the cataclysmic climate fluctuations of the mass extinctions. The runoff from melting glaciers inundated rivers, raising ocean levels at least 300 feet. In northern Asia, the top layers of soil, frozen for tens of thousands of years, began to melt during the warmer summers. The fertile mammoth grounds between modern England and the Netherlands became the North Sea.

"I think there's only one possible explanation for the extinctions: the conditions during the Pleistocene-Holocene transition were different from those at the end of the previous ice ages," says Andrei Sher. He and other "overchill" advocates such as Valentina Ukraintseva of the Komarov Botanical Institute in St. Petersburg and R. Dale Guthrie of the University of Alaska argue that the shifting weather patterns, which brought more rain and snow, led to the mammoth's demise, particularly in Siberia. The idea is that across the High Arctic, the cold, dry grassland—the mammoth steppe, as Guthrie dubbed it—was transformed into boggy tundra that produced little fodder in summer and was buried under snowdrifts in winter. The grasses and herbs of the tundra were supplanted by less nutritious mosses and sedges. Populations of mammoths and other large herbivores dwindled as their habitat disappeared.

Another important piece of evidence for overchill is the physiological change in the creatures themselves during the late Pleistocene: mammoths in northern Siberia grew smaller as the Ice Age drew to a close. As early as 1908, Konstantin Vollosovich had speculated that small mammoth bones found in the northern reaches of Siberia were a sign of a species on the way out—the beasts were not getting enough to eat to grow to their full size. This might not have happened due to the classic overchill scenario, contends Sergei Vartanyan, a mammoth expert in St. Petersburg. He believes that the mammoth steppe became fragmented as warmer temperatures allowed the forest taiga to advance north. The forest penned in the mammoths, which preferred the open steppe, forcing them to compete with one another for diminishing forage and thus curtailing the growth of individuals. Dwindling food supplies eventually killed off the species, Vartanyan argues.

To make their argument more persuasive, overchill adherents need more specimens of frozen mammoths or mammoth bones indicating that the animals were suffering from malnutrition or attaining shorter statures at the end of the Great Ice Age. And they would have to tie this to evidence that vegetation in the region where the animals died had changed from grasses or other healthful mammoth feed to less nutritious sedges and mosses. That's a tall order—so tall, in fact, that many overchill enthusiasts are pessimistic about making their case in North America, where there's virtually no evidence that mammoths were suffering from malnutrition at the end of the Pleistocene. In Eurasia, on the other hand, the findings of Vartanyan and others suggest that overchill has a chance.

— — —

But it's a slim chance. Challenging the neat picture of plant succession meting out a fatal blow are fossil finds suggesting that many mammoth populations across Eurasia were dying out long before the end of the Great Ice Age. The mammoth had disappeared from much of Europe, it seems, before losing ground in Asia and North America. Also countering the climate hypothesis, some argue, are the tusks. The biggest teeth in the animal kingdom, tusks, like trees, grow in layers. "They form deep within the skull and build up like a stack of cones," explains Daniel Fisher, a paleontologist with the University of Michigan. And, like tree rings, the cones of a tusk reveal information about the age, nutrition, and environmental conditions of the beast, even its health from week to week. According to Fisher, the thick tusks from mammoths and American mastodons that died near the end of the Pleistocene suggest that the animals "were growing as fast and maturing as young as any elephants alive today." That denies the notion of starving mammoths.

Pavel Putshkov, a zoologist in Kiev, Ukraine, vehemently opposes the overchill hypothesis, claiming that mammoths were far too malleable in their dietary habits and too resilient to bad weather to succumb to climate change. In his witty 1997 treatise "Were the Mammoths Killed by the Warming?" he argued that it is wrong to assume that woolly mammoths were so dependent on the dry, cold conditions of the steppe tundra that they could not survive in other habitats. He pointed to ample evidence suggesting that mammoths lived in the forests of the central Russian plains at the height of the Great Ice Age, noting also that their fossils have been unearthed in alpine valleys in the Caucasus and the southern Crimea.

He also finds plausible a controversial hypothesis that the best habitat for the woolly mammoth was Arctida, a land envisioned by the Russian permafrost engineer Savelii Tomirdiaro to have lain north of Siberia during the Pleistocene. Based on years of observations in the New Siberian Islands in the Arctic Ocean, Tomirdiaro argues that over millennia, wind-blown silt accumulated on permanently frozen sea ice in the Arctic, like dirt piling up on a frozen lake. Eventually grasses took root in the silt, and Arctida became a vast, treeless meadow. The landmass, Tomirdiaro contended, stretched from at least the Laptev Sea, off Siberia's northern coast, to the Bering Sea, forming a much broader connection between North America and Eurasia than the Bering Land Bridge.

Although the extensive ice cover might have allowed winter temperatures to drop as low as −100 degrees Celsius, summers must have been reasonably warm.

Plant remains found in Pleistocene deposits in the New Siberian Islands suggest that Arctida, if it existed, was indeed grassy and more humid than the dry steppes on the mainland. Judging from the numerous mammoth remains found on the New Siberian Islands, the animals thrived in this Arctic paradise—at least during the summer. In winter at those latitudes, asserts Paul Martin, the mammoths "must have just stood around for months. They couldn't afford to shiver, or they would lose too much energy and die fast. They would have had to practically go into hibernation. Maybe they were good at meditation."

Master of a vast territory—from Arctidian meadows to the Crimean highlands and central European forests and on to the North American plains—the woolly mammoth, Putshkov says, should have been far less susceptible to the ravages of climate change than were limited-range species, like the reindeer that survived the Pleistocene-Holocene transition. Studies of African parks under ecological stress support the idea of the mammoth as a survivor. During severe droughts in Kenya's Tsavo National Park in 1960–1961 and 1970–1971, elephants denuded the land with their rapacious appetites. But their wide-ranging tastes—from roots and bulbs up to the crowns of trees—and their ability to use their tusks to dig wells in dry lakebeds and puncture baobob trees for water allowed them to hold out longer than other animals in the park. There's no reason to assume the mammoth was more finicky or less resourceful than the elephant.

Putshkov also derides the view that the mammoth could not cope with the wetter winters of the Holocene. Heavy snowfall might slow a mammoth, he argues, but only in rare cases would the snow accumulate to life-threatening depths. After all, if reindeer could survive, why not the longer-legged mammoth? Some experts have argued that mammoths exerted far more force per square inch when they walked than reindeers exert, which would make them prone to becoming mired. Pads on the bottom of their feet reduced the pressure, but their biggest advantage, according to Putshkov, was their ability to power through snow like a truck. The mammoth, he argued, "was a living cross-country vehicle that marched on any ground more surely than other herbivores!"

How they managed to eat after heavy snowfalls is a puzzle, although Putshkov believes they could have used their tusks to break through icy crusts, then reached through the snow with their trunks to the grass below. He doesn't doubt that mammoths browsed on twigs and bark when necessary.

If the going got really bad, some argue, mammoths would have migrated to more temperate regions. The explorer and wag Bassett Digby was sure of it. The notion that mammoths would not undertake seasonal migrations—that "the mammoth was too big a fool to walk after its southward receding food"—was absurd, he wrote in 1923. He gave the mammoth more credit.

The idea of mammoths trudging hundreds of miles for better forage remains pure speculation. The first attempt to substantiate the idea by analyzing a mineral (strontium) in the animal's teeth failed. The ratio of strontium isotopes in the land varies from place to place, and the accumulation in teeth provides a good marker of the region in which an animal fed. In 1999 scientists examined strontium levels in molars from fifty-eight mammoths and mastodons that had died in Florida. Their findings suggested that the mastodons did migrate north into Georgia, but the mammoths then stayed put. Larger studies in other regions will have to probe this question further, but advocates of the overchill hypothesis argue that even if migration can be proven, the extreme swings of climate would have eventually brought the peripatetic beasts to their knees.

But the climate was not the only thing changing as the Great Ice Age wound down. Waves of prehistoric hunters were invading the New World in the twilight of the Pleistocene, venturing from Asia into North America, across an isthmus that was bounded to the south by the mountains of the present Aleutian Islands. To overchill proponents, these prehistoric hunters were finishing off the last mammoths of a dying breed. Another group, however, contends that the hunters encountered thriving herds of mammoths in the New World and went on a killing spree.

KILLER WAVE OF THE NEW WORLD

Not satisfied with the idea that colder, wetter winters could wipe out a species as magnificent as the mammoth, a few scientists in the mid-1960s began wondering if a predator could be to blame—and found themselves looking in the mirror. Overhunting, they argued, would explain why mammoths died out gradually across Europe and Asia, appearing to beat a retreat into northern Siberia—perhaps in fear of hunters—before making their last stand on Wrangel Island in northeastern Siberia 3,700 years ago. But little more than the pattern of mammoth extinctions, a slow but steady decline from west to east, has been found as evidence against humans throughout Eurasia—certainly not enough evidence to convict our ancestors. However, a more persuasive case can be made in North America, the destination of bands of hunters that crossed the Bering Land Bridge between 12,000 and 15,000 years ago.

Whether they were intrepid explorers or unlucky groups

separated from kin by the rising ocean levels that submerged the isthmus's shrubby tundra, these early Americans did not stay long on Beringia, the land at the top of the world. As the glaciers retreated deeper into modern Canada toward the end of the Great Ice Age, the hunters swept down a finger of land between two ice sheets, then fanned out across the North American plains. They brought a skill: how to chip spear points from flint. These distinctive fluted points, 2–7 inches long and shaped like a compressed arch, have turned up all across the New World since their discovery in 1932. They are named after a particularly rich trove of points found in Clovis, New Mexico.

Armed with spears and eventually the atlatl, a stick that slings spears at high speed, the Clovis hunters had plenty of big game to stalk—including mammoth—and no real foes. Based on a handful of kill sites where Clovis points have turned up with mammoth bones, it appears that the hunters liked to ambush their prey near water. The Clovis hunters did not limit themselves to mammoths, which must have been riskier to bring down than smaller game. Recovered bones reveal a diverse menu, from mastodon and bison to black bear and horse.

Indeed, how they felled their huge quarry is a puzzle. A recent experiment to use simulated Clovis weaponry to penetrate the rubbery skin of dead zoo elephants for the most part failed, except when the points were driven into a few vulnerable spots in the hide. But these creatures vanished from North America soon after the Clovis arrived. Was it a coincidence?

— — —

Paul Martin was a young ecologist when, in 1967, he argued that the North American animals lost at the end of the Great

Ice Age were those one would expect to lose if they had been hunted. The prehistoric hunters, he imagined, would have feasted on big, plodding herbivores: mammoths, mastodons, giant ground sloths, and camels. He had no trouble placing hunters at the crime scene. In as little as 350 years after crossing the Bering Land Bridge, Clovis hunters penetrated to the Gulf of Mexico, and within a millennium they reached the southern tip of South America. Clovis hunters may not have been the first to people on North American soil—scientists hotly debate the evidence of settlers before Clovis from Asia and Europe as early as 50,000 years ago. But if there were North Americans when the Arctic hunters crossed the land bridge and turned south, it appears they were scarce and clung to the coasts.

This is important because large numbers of pre-Clovis peoples in the New World would undermine Martin's hypothesis, which relies on the mammoths and other quarry being unfamiliar to humans—and thus not wary of hunters—when the Clovis people arrived. But were there enough Clovis hunters to kill off so many species? After all, the continent lost three out of every four of its megafauna species in a few centuries. Six years after his initial call to the scientific community, Martin argued in a classic paper in *Science,* "The Discovery of America," that the Clovis hunters were like a shock wave spreading across the continent. The highest population densities were at the wave's bow, he wrote, where small groups wreaked concentrated havoc.

Clovis hunters thrived on the unspoiled continent, proposed Martin, who speculated that their population might have doubled every 50 years or less. Fueling this explosive growth were ungainly game that had no fear of humans. "The hunters would not have needed to plan elaborate cliff drives or to build clever traps"—strategies that Native Americans used until the nineteenth century to hunt bison. According to Martin, the Clovis slaughtered the naïve beasts, taking less than a decade on average to wipe out big game from a given locale. "Unless one insists on believing that Paleolithic invaders lost enthusiasm for the hunt and rapidly became vegetarians as they moved south from Beringia, or that they knew and practiced a sophisticated, sustained yield harvest of their prey," he wrote, "one would have no difficulty predicting the swift extermination of the more conspicuous native American large mammals."

Toward the end of the killing spree, with many species in decline, the human population itself must have crashed. Martin imagined a harmonious confederation of hunters fracturing into tribes that quarreled over shrinking hunting grounds. With the days of easy kills gone, the Clovis would have had to subsist on roots and nuts, eventually learning how to cultivate favorite plants—the beginning of North American agriculture.

Clouding the elegant scenario, however, is the scarcity of supporting physical evidence. While fluted points are common across North America, at only fifteen locations have they been found alongside mammoth bones that appear to have been scraped or gouged by people—too few for a continental slaughter, critics say. More frustrating to backers of the overkill hypothesis is the dearth of evidence linking Clovis hunters to prey other than mammoths. Besides a few sites with Clovis points and mastodon bones, nowhere have archaeologists found signs of early Americans hunting other large animals, like ground sloths or camels, that also went extinct at the end of the Great Ice Age.

Although he retired in 1989, Martin maintains an office at the Desert Laboratory, a collection of rustic, slate-roofed buildings with thick walls of andesitic basalt founded in 1903

Above: At the turn of the 19th century, a merchant in the Siberian outpost of Yakutsk made this fanciful drawing of the Adams mammoth based on secondhand descriptions of the beast's carcass (*Courtesy of the Mammoth Committee of the Russian Academy of Sciences*)

Below: Otto Herz and Eugen Pfizenmayer, standing on either side of their homemade mammoth flag, along with Lamut hunters who built a cabin for the team's fall 1901 excavation of the Berezovka mammoth (*Courtesy of the Mammoth Committee of the Russian Academy of Sciences*)

Above: The Berezovka mammoth remains made part of the journey from Siberia to St. Petersburg by horse-drawn sledge (*Courtesy of the Mammoth Committee of the Russian Academy of Sciences*)

Below: The stuffed Berezovka mammoth, in its presumed death pose, is on display at the Museum of Zoology in St. Petersburg (*Courtesy of the Mammoth Committee of the Russian Academy of Sciences*)

Siberian gold miner Anatoly Logachev received a watch for having reported the discovery in 1977 of the baby mammoth, Dima. The Soviet government later insured this specimen—the most intact mammoth ever found—for 10 million rubles (*Courtesy of the Mammoth Committee of the Russian Academy of Sciences*)

In the early 1970s, Soviet scientists excavated nearly 9000 mammoth bones from a bend in Siberia's Berelekh River, the most prolific mammoth site ever discovered (*Courtesy of Nikolai K. Vereshchagin*)

Akira Iritani hopes to resurrect the mammoth through cloning, using a technique he helped pioneer: intracytoplasmic sperm injection; equipment for the technique is shown in photo (*Courtesy of Mutsumi Stone*)

Kazufumi Goto is leading the hunt for mammoth sperm, which he intends to inject into an elephant egg to create a hybrid female. Subsequent generations would look increasingly mammoth-like (*Courtesy of Mutsumi Stone*)

In 1999, Goto's team found a piece of ancient skin—apparently from a woolly rhino—in Pleistocene sediments in Siberia (*Courtesy of Mutsumi Stone*)

In October 1999, an expedition led by Bernard Buigues hauled a 23-ton block of frozen Siberian ground to an ice cave for study. The team had reason to suspect that the block contained an intact woolly mammoth *(Courtesy of Francis Latreille/Cerpolex)*

Since the end of the Soviet Union, scores of indigenous Siberian Dolgans have returned to a nomadic life on the tundra, living out of reindeer-drawn mobile homes called baloks *(Courtesy of Francis Latreille/Cerpolex)*

Above: In expeditions on the Taimyr Peninsula in 2000, Buigues's team harvested from the tundra exactly 100 mammoth tusks. Information on local climate and the health of individual mammoths during their lifetimes can be gleaned from the tusks, which are deciphered similarly to tree rings (*Courtesy of Tim DeClaire/Cerpolex*)

Left: In Khatanga in October 2000, Dick Mol holds a mammoth bone in place as Ross MacPhee drills into it to take a sample for DNA analysis (*Courtesy of the author*)

Above: MacPhee began his hunt for a hyperdisease pathogen in summer 1998 on Wrangel Island, famed as the stomping grounds of the last known mammoths 3700 years ago (*Courtesy of Clare Flemming*)

Right: Sergei Zimov is the mastermind behind Pleistocene Park, an experiment in northeastern Siberia that is attempting to transform boggy tundra into fertile grassland reminiscent of the mammoth steppe (*Courtesy of Mutsumi Stone*)

Bernard Buigues and his team in October 2000 using hair dryers to defrost the top several inches of the permafrost block containing the remains of the Jarkov mammoth (*Courtesy of the author*)

Buigues inspects the exposed vertebrae of the Hook mammoth, which he and his team excavated in April 2001 (*Courtesy of the author*)

on Tumamoc Hill, above Tucson, Arizona. Petroglyphs scattered across the hilltop are relics of a fortress kept here 900 years ago by the Hohokam. Today, the butte dotted with saguaro cactus is the home of a 400-hectare reserve managed jointly by the University of Arizona and the U.S. Geological Survey. In his office, next to an antique roll-top desk, is a glass case displaying an extinct ground sloth's fossilized dung. It's a softball-size relic that Martin found in a cave in the early 1980s in the Glen Canyon Recreation Area. He would trade that dung in a heartbeat for news of an archaeological site with both Clovis points and the bones of butchered ground sloths.

While Martin rues the dearth of archaeological evidence for his hypothesis, he does not consider it a fatal flaw. Other facts fit the picture well, he says. For example, small animals on continents were not killed off, nor were marine mammals —"They go through the extinction event without a whimper," he says. Nor is he surprised that there are so few sites. For one thing, he says, the Clovis people, advancing rapidly across the continent, didn't tarry long enough in any one place to accumulate vast piles of bones. Kill sites rarely persist in the fossil record, even in Eurasia, where people and mammoths commingled for millennia. And no bones have been found at verified kill sites in Africa, where elephants were systematically butchered in the nineteenth century to clear land for agriculture or killed for their ivory, then in high demand for piano keys.

Like-minded colleagues have sought to modify Martin's overkill hypothesis to better fit the available data. Larry Agenbroad, a mammoth expert at Northern Arizona University, for instance, doubts that the spear-toting Clovis hunters had enough power and efficiency to purge an entire continent filled with big game. Instead, he proposes, mammoth populations were not evenly distributed across North America, and that the most likely route of southern migration from Beringia brought the Clovis hunters smack into the heart of the animal's range, in the modern U.S. Southwest. The hunters enjoyed a few decades of easy kills before chasing the survivors into less suitable habitats. "The die-off then spread concentrically, much like the ripples created when a stone is dropped in a pond," he says.

While Martin doesn't put much stock in the overchill hypothesis, other overkill proponents aren't so quick to discount the effects of climate; they simply claim that weather patterns at the end of the Pleistocene abetted the main culprit. During a crucial period when hunters were presumably inflicting the heaviest losses on mammoth herds, from 10,900 to 11,300 years ago, the western United States was parched by drought, according to evidence gathered by the anthropologist C. Vance Haynes of the University of Arizona. In times of stress, mammoths, like elephants, likely would have congregated near springs, leaving themselves open to Clovis ambushes.

Overkill critics have questioned why species like the bison managed to survive this onslaught. The answer might lie in how quickly a population can replenish itself. A mature female bison can have a calf every year, while cow elephants—and presumably mammoths—give birth only every four to six years. "If you wipe out young animals," says Agenbroad, "pretty soon you have a problem with the viability of the population."

Considering the sheer number of species that went extinct, that argument runs thin—even if Clovis men were hunting from morning until night. "You have to hunt a mammoth, you have to hunt a camel, you have to hunt a horse. If you don't have a rifle, that's a lot of work. It would take them days to hunt and kill a mammoth," says the Dutch Pleistocene expert John de Vos, who believes strongly that climate change, and the associated turnover of plant species, doomed the mammoth and its contemporaries. The overkill idea certainly doesn't hold water in Europe, he contends, pointing out that among the several thousand mammoth bones he oversees at the National Museum of Natural History in Leiden, "not a single one was worked by man." Even where there's convincing evidence that prehistoric people ate mammoths—witness the site in Poland with the discarded pile of mammoth tongue bones—it's often impossible to tell whether humans hunted the beasts or scavenged those already dying or dead.

One attractive proposal is that overhunting was like unraveling a fabric by pulling a single thread. With mammoths such a dominant, omnipresent player across the Northern Hemisphere, Putshkov wondered what would happen if only they had been overhunted and other species spared. He concluded that mammoths, along with the less abundant woolly rhinos, must have been "mega-gardeners," essentially dictating the makeup of the plant communities in their habitats by grazing, trampling, and fertilizing the ground with their dung and urine. Smaller herbivores relied on these mega-gardeners to maintain the habitats thus, while large predators like cave lions kept the woollies in balance by culling weaker animals.

When humans killed off a large enough proportion of mammoths, the environment literally fell to pieces, Putshkov argues. Because the mammoth was vital to the health of its ecosystem, removing it led to the deterioration of habitats and the demise of many mammoth contemporaries. That explanation would hold up for all of the regions from which the mammoth disappeared, not just North America, he asserts.

Kenya's Tsavo National Park again buttresses this argument. As elephants were hunted to the verge of annihilation in western Kenya before the eighteenth century, the savanna was gradually supplanted by thorny bushland—without any corresponding climate change. Wildlife went into a tailspin. Only after the park was established in 1948, when elephants gained a measure of protection, did the East African megafauna make a comeback in the region.

Although mammoth populations surely fluctuated with the severe climate changes marking the transitions between the glacial advances and retreats during the Pleistocene, the species succumbed only after the Great Ice Age. In Putshkov's view, the sole factor setting the Ice Age apart from previous glacial advances was that prehistoric humans had acquired technology—more sophisticated weapons, social organization, and fire control—that allowed them both to defend themselves from large predators and to slaughter mammoths.

Strengthening the argument are a handful of large animals either presumably hunted to extinction (like the flightless moas of New Zealand) or brought to the brink before the slaughter was stopped (like the plains bison of North America). If prehistoric New Zealanders could obliterate the moas, which were abundant on the islands before humans arrived, then perhaps Clovis hunters could wreak similar havoc on naïve populations of Columbian and woolly mammoths in North America. In northern Eurasia, on the other hand, woolly mammoths may have already been declining by the time hunters arrived to deliver a coup de grâce. The well-made case of overhunting to extinction that most resembles this scenario is the demise of the Stellar's sea cow.

In 1724, near the end of his ambitious explorations, Peter the Great appointed Vitus Bering, a Danish officer in the Russian navy, to find out whether the continents of Asia and America were connected. Peter died a month later, but Bering continued his expedition. He spent the next six years in eastern Siberia, mapping the territory, but failed to answer the question. He got a second chance in June 1741, this time sailing east from the Pacific port of Okhotsk on the *St. Peter.* His ship's doctor was Georg Wilhelm Steller, a young German naturalist who had come to the eastern fringe of Siberia seeking adventure. Six weeks after setting sail, Bering landed on the Alaskan coast. The continents, it appeared, were indeed separated by water. Going ashore, Steller discovered several species—including a type of blue jay that bears his name. Later that day, however, Bering ordered Steller back to the ship: they were returning home. In "The Auk, the Dodo, and the Oryx," Robert Silverberg recounts the sudden end to the expedition: "Steller was outraged at this turn of events. He had envisioned a leisurely exploration of the new land. But Bering, an old and tired man, had lost his stomach for discovery. He had done what the government had asked him to do: he had crossed the strait east of Siberia and had found the western shores of America. Now he was impatient to return."

The return voyage was plagued by trouble. Many sailors came down with scurvy, and high winds forced Bering to tack northward. By early November twelve men had died, and Bering, still hundreds of miles from Russia, decided to lay up near an island for the winter. He himself was ill and died that December, a few days after the *St. Peter,* anchored in a cove, was wrecked in a storm.

By spring, many of the castaways had recovered, thanks to the creatures they called mermaids. These were sea cows, 30-foot-long cousins of the Florida manatee and the dugong of Southeast Asia and distant relatives of the mammoth. Growing tired of the meat of sea lions, Steller and his companions tried hunting the whiskered mermaids. After they finally landed one with a harpoon, Steller made this heartwrenching observation: "We ... noted, not without astonishment, that a male came two days in succession to its female which was lying dead on the beach, as if he would inform himself about her condition." Killing a single sea cow provided enough meat, which tasted like beef, to last for weeks.

The sea cow saved the castaways, which in turn doomed the sea cow. That summer the crew built a new ship from the wreck of the *St. Peter* and by August set sail for Kamchatka, arriving in the harbor of Petropavlovsk. Four years later, Steller died after a brief illness at the age of thirty-seven.

Word of the life-saving sea cow got around, and Russian fur traders plying the waters between Russia and Alaska would often stop at Bering Island to supply their ships with sea cow meat. Less than 30 years after Steller and his mates had survived a winter on Bering Island, the fur traders had hunted the mermaids into oblivion. The last recorded kill occurred in 1768; all that was left was bones.

NASTIER THAN EBOLA

Ross MacPhee has spent much of his career studying mammalian extinctions in the West Indies, puzzling over mysteries largely passed over by other scientists. In the early 1990s, he began to focus on the extinctions since 1500, the beginning of Europe's exploration and colonization of the New World. Like the overkill advocates, he too has found a correlation between animal extinction and the arrival of humans; however, he draws a very different conclusion.

When MacPhee, the curator of mammals at the American Museum of Natural History, and his colleague Clare Flemming compiled a list of species worldwide that have disappeared in the last five centuries, only one in every eight extinctions were so-called megafaunal species. This proportion was much different at the end of the Great Ice Age, when three of every four mammals that went extinct in the continental New World were megafaunal. And in the 10,000 years between the end of the Ice Age and the start of

European exploration of the New World, no more than one or two megafaunal species vanished from continental North and South America. "Nobody ever talked about this hiatus," MacPhee says, "and that led me to wonder what was actually going on." Nobody knew. "Evolutionary biologists tend to think about how species arise," he says. "They don't think about the end."

Yet despite the differences, one recurring theme of both the prehistoric extinctions and the losses in the last five centuries, he noted, was a first encounter between people and the soon-to-be vanquished mammals. It happened on nearly all habitable landmasses except Africa and Eurasia, where humans evolved alongside animals and relatively few extinctions occurred in the past 40,000 years. MacPhee wrestled with a conundrum: "Why is it that again and again, all over the planet, you find the same story: that there's really nothing happening until people come, and then the animals go down?"

MacPhee began to think about these sorts of questions as a student on an archaeological project to search for evidence of the earliest Americans. Born in Edinburgh, he and his family moved to Edmonton when he was three. During his freshman year at the University of Alberta in 1967, he joined his first dig. "I really knew nothing about paleontology or archaeology. And I didn't like it, I wasn't patient enough." But the scientific process—formulating an idea and testing it—left an indelible impression. After graduating he got a doctorate in biological anthropology. He taught gross anatomy for 10 years at Duke Medical School before applying for the mammal curatorship that had opened up at the museum. To his surprise he got the job, and he began to carve out a niche in the scientific community through his work in the West Indies.

MacPhee, with close-cropped brown hair, a trim white beard, and glasses, is an avowed skeptic of the two leading mammoth extinction theories: that mammoths succumbed to either human hunters or habitat degradation from climate change at the end of the Pleistocene Epoch. He even rejects attempts to reconcile these ideas through the notion that our ancestors applied a coup de grâce to a dying breed. While overkill is "a very compelling argument," MacPhee says, he doesn't believe it. "The whole notion of big-game hunters madly dashing around the landscape, killing everything in sight, is completely inconsistent with the anthropological picture," he says. The ascendant view in North American archaeology is that it was female gatherers who kept their clans well fed. To mete out as much punishment on mammoth herds as Martin and other overkill adherents suggest, MacPhee claims, "the hunters must have been total Rambos." And that, he says, is highly unlikely. To support his argument he points to Africa, where much higher densities of people, armed with guns and able to slay elephants for sport or ivory at will, failed to eliminate the species before protective laws were enacted in the last century. (Of course, one could argue that the human hunters did not have enough time to finish off the species before the laws were passed.)

While mammoths may have trusted people naïvely on first contact, MacPhee doubts that they would have remained blind to the threat of the two-legged interloper for long. Even if humans regularly hunted mammoths, he believes the species would have withstood the pressure. Apart from island-bound species, animals have proven resilient to rapacious predators. Two centuries of intense whaling, for example, while coming perilously close, has failed to annihilate a single species.

Even naïve populations can quickly acquire survival skills.

A recent study found that moose living without wolves for as few as 50 years were easy prey when wolves were reintroduced into their territory—essentially experiencing Martin's blitzkrieg. Within a generation, however, the moose population had learned to key in to wolf howls and scents. In their report in the February 9, 2001, issue of *Science,* the authors speculated that "perhaps species that failed to survive post-Pleistocene hunters were simply not quick learners." It would require a leap of faith, however, to believe that mammoths were not like elephants, which do learn quickly.

MacPhee's Eureka moment came in 1992 after reading a gripping article in the *New Yorker* describing fears of a potential outbreak of the deadly Ebola virus in suburban Washington, D.C., and how the pathogen has flared up on occasion to savage villages in Africa. Sometime in the 1980s, the virus appears to have leapt from an unidentified species, perhaps a bat, into humans. There's no known treatment or vaccine to protect people against Ebola, which disintegrates organs and blood vessels. The worst strain, Ebola Zaire, kills nine of every ten people infected. "The idea that a disease could literally burn through a population was mind-boggling," MacPhee says. He began to wonder whether a pathogen could become powerful enough to drive to extinction not just one species but several, including the mammoth and other large mammals of North America at the end of the Great Ice Age. He raised the idea with Preston Marx, a virologist at the Aaron Diamond AIDS Research Center in New York.

Marx had trekked to places like Gabon and Cameroon to sample green monkeys, kept as pets or shot for the bush meat market, in search of the primate version of the AIDS virus. "It was clear to me that this was the kind of guy who's willing to go out on extreme limbs to investigate things," says MacPhee.

"So I asked him: 'What do you think of disease as a cause of extinction?'" Marx listened, intrigued, before rattling off a host of problems that had to be overcome for the idea to blossom. He doubted, for instance, whether a single pathogen, no matter how vicious, could annihilate species as diverse as mammoths and camels, woolly rhinos and sabertooth cats, while allowing others—like the reindeer and the bison—to escape relatively unscathed. "But he was interested enough to give it a shot," MacPhee says.

Presenting their hyperdisease idea at a 1997 meeting on extinctions on Madagascar, they encountered considerable skepticism. But few scientists were willing to dismiss the idea out of hand. "It's worth giving its fair day," says the overkill guru Paul Martin. "It's made me look at the literature more critically about what disease does and how it jumps. Maybe disease could be at least part of the story."

For a pathogen to qualify for hyperdisease status, it must, like Ebola, kill the majority of its victims (at least 75 percent, Marx estimates) within days. Young adults and offspring must succumb to the disease, destroying a population's ability to reproduce. "You've got to have something really nasty," MacPhee says. The pathogen must also cross species lines with ease, perhaps using a "burning bridge strategy," meaning a microbe that spreads quickly as it kills off its hosts.

In today's gallery of rogues, no bug rises to hyperdisease status—although some come close. The Spanish flu, unlike other strains known through history, had a high mortality rate in young adults and wiped out whole Eskimo villages as it claimed 40 million lives worldwide in 1918. This strain clearly jumped the species barrier: it seems to have been the same bug that killed millions of pigs that year, and researchers were able to infect and cause disease in ferrets—although attempts to induce illness in rats, monkeys, and many other lab animals failed. Ebola, meanwhile, would seem another possible candidate: even the less-virulent strain, Ebola Sudan, easily kills three out of four people it infects. But the virus is known to be lethal only to primates. And while various strains of the mycobacterium that causes tuberculosis will infect humans, bison, cheetahs, lions, and other mammals, it hardly kills fast enough to endanger a species.

One recent experience suggests that disease can conspire with other factors, such as habitat loss, to drive a species to extinction. In a nightmarish incident on the Hawaiian island of Maui in the 1820s, an American ship arriving from Mexico dumped water contaminated with the larvae of *Culex quinquefasciatus,* the mosquito that carries avian pox and malaria, into the harbor. By the early 1900s, many native Hawaiian songbirds had retreated into the mountains, often into much poorer habitat, but away from the diseased communities hugging the shore. For some, the losses were too heavy to sustain: a sharp-beaked black bird called the Kauai o'o hasn't been seen in years, and a brown bird with a creamy tan belly, the po'ouli, is down to its last few individuals and is expected to soon go extinct.

More recently, a superbug may have terrorized the amphibian world. In the early 1990s, a fungus slipped into populations of the golden toad in Costa Rica and Panama and went on a rampage, causing a suffocating skin lesion. No golden toads have been observed anywhere since 1995, suggesting that the species is now extinct. "This was a novel pathogen and essentially an overnight collapse," MacPhee says. "That's what a hyperdisease would be like."

~ ~ ~

Considering that the Clovis people might have taken only a few centuries to penetrate North America, it's conceivable that they or their dogs could have been the messengers of death—not as hunters, but as carriers of a hyperdisease pathogen with which they inoculated most of the mammoth population. MacPhee and Marx can only speculate about how such a transmission could have occurred. Perhaps when stalking mammoths, the hunters got near enough that a microbe transmissible by air was passed along to the beasts. Or perhaps an intermediary carrier, a rodent, might have transmitted the bug from humans to mammoths. Historically, rats have played such nefarious roles, helping to spread the bubonic plague from China to Italy and the rest of Europe in the mid-1300s. Within five years, the Black Death had killed 25 million Europeans—one of every three on the continent. According to the fourteenth-century Italian writer Giovanni Boccaccio, some victims became ill and died so quickly that they "ate lunch with their friends and dinner with their ancestors in paradise."

Once infected with a hyperdisease pathogen, mammoths could have easily spread the disease; indeed, mammoths might have been particularly vulnerable to extirpation. If mammoths, like elephants, could have a single calf only every five or six years, mature females resistant to the pathogen might not have been able to breed a replacement generation fast enough to save the species. Smaller mammals that breed faster would have stood a greater chance of surviving. Meanwhile, isolated groups that survived the plague could have succumbed to inbreeding or other blights brought on by a shrunken gene pool.

If such a superpathogen did exist, it may have died out with the mammoths and other victims. Or it may persist as a kind of mammoth version of smallpox. The Variola virus, which causes smallpox, was a leading cause of death before a global vaccination campaign eradicated it from the wild; a close cousin, varicella zoster, triggers chicken pox, a mild illness if contracted in childhood. Could some pathogen today that poses little threat to us—perhaps causing something like chicken pox—have been a terror to mammoths, infecting them with something as deadly as smallpox?

In my opinion, intriguing circumstantial evidence for a plague can be found near Sevsk, a town 250 miles southwest of Moscow. During the late 1980s, scientists from the Institute of Paleontology in Moscow unearthed some 4,000 bones from thirty-six woolly mammoths, including seven full skeletons, from a sandy pit outside Sevsk. They found the remains of young and old alike—from a six-week-old baby, its tusks just beginning to protrude from its cheeks, to a fifty-year-old bull. There's no evidence that hunters felled these mammoths, which died about 13,680 years ago, nearly the end of the line for mammoths this far south in Russia. What calamity befell them remains a mystery; perhaps they all plunged into a sinkhole—or perhaps they were all killed by hyperdisease.

But Sevsk is in southern Russia, far from the boundary where the ground is permanently frozen. Thus the Sevsk bones are fossils: minerals have replaced most of the mammoth's original bone tissue, driving out evidence of pathogens as well.

MacPhee knew that his chilly reception at the Madagascar meeting was largely due to lack of evidence. And he knew his best hope for finding a superpathogen lay in frozen mammoth bones or other tissues. "The fact that there are these frozen Ice Age beasts still around means we can test the hyperdisease theory in an empirical way," he says. In North America, only a few northern patches in Alaska and Canada have permanently frozen ground that may have preserved fragile DNA or cells from so long ago. Most of Canada was buried under glaciers during the Pleistocene, while the prime habitat for megafauna to the south hasn't stayed cold enough to prevent most remains from deteriorating during the modern Holocene Epoch—except high in the Rocky Mountains, perhaps, where the megafauna did not live.

So MacPhee set his sights on northern Siberia, where millions of acres of Pleistocene sediments, rich in the remains of mammoths and other extinct animals, have lain frozen since they died. The youngest mammoth bones in the world come from remote Wrangel Island in the East Siberian Sea, high above the Arctic Circle. The mammoth made its last stand on Wrangel around 3,700 years ago, 700 years after the Egyptians built their great pyramids at Giza. The earliest evidence of people on the island, stone tools and spear points, is about 3,100 years old. Slightly older human artifacts or slightly younger mammoth bones make it conceivable that the hunters arrived when the mammoths were still alive. MacPhee knew that the only way to find out whether the first humans on the island brought a killer plague was to go to Wrangel and retrieve the frozen bones and other tissues of the last mammoths to have walked the earth.

~ ~ ~

There's a possibility, however remote, that instead of finding a villain that killed off the mammoths, MacPhee—or any other mammoth hunter—might stumble on a pathogen that is dangerous to us. The Siberian tundra, and the Pleistocene mummies it holds, could be incubating the next horrific human plague.

We can only guess at the bacteria or viruses lurking in the cells of creatures long dead. "We might not know what these pathogens look like," warns John Critser, the scientific director of the Cryobiology Research Institute at the Indiana University Medical Center. Mammoth sperm injected into an elephant egg might bring with them an unwelcome surprise: ancient forms of deadly modern diseases known to be transmitted when the donor animal's sperm is used in artificial insemination, including viral diarrhea, hepatitis B, tuberculosis, salmonella, and swine fever. Scientists will have to screen carefully for pathogens before transferring the essence of mammoth into an elephant, says Critser, who has mulled over this issue in the course of his work transplanting elephant ovarian tissue into mice to produce elephant eggs. "There's a very real potential of introducing infectious diseases in elephants."

Some pathogens survive being flash frozen in liquid nitrogen, an unimaginably cold minus 190 degrees Celsius. And microbes can live for hundreds of thousands of years in the permafrost, as David Gilichinsky of the Institute of Physicochemical and Biological Problems in Soil Science in Pushchino, Russia, has shown in his pioneering studies on the tundra.

A recent expedition wrapped up on a brisk summer day at Point Chukochii, in northeastern Siberia, as a member of Gilichinsky's team yanked threaded steel rods out of a hole in the tundra like a magician laboriously pulling one knotted handkerchief after another from a hat. The sun had broken through after several days of rain and snow, but even in August, drilling for microbes high above the Arctic Circle is no picnic. The hunk of metal that emerged from a depth of nearly 100 feet revealed a tip bearing a plug of what looked, unremarkably, like frozen dirt. In fact, the plug harbored millions of microbes that, once liberated from their Siberian prison, resumed normal activity. This chunk of wind-blown sediment, deposited on the shores of Lake Yakutskoe 40,000 years ago and frozen ever since, was alive.

An early hint that permafrost may not be a sterile wasteland came in 1911, when Russian researchers reported that they had cultured benign bacteria from the Berezovka mammoth. Although many scientists now suspect that modern bacteria had invaded the carcass, those findings—along with indications from later research that ancient permafrost soil may contain viable life—intrigued Gilichinsky. In 1979, with the microbiologists Dmitrii Zvaygintsev and Elena Vorobyova of Moscow State University, he began hunting for microbes near the spot where Russia's Kolyma River empties into the East Siberian Sea. His team quickly tapped a microscopic menagerie of bacteria, fungi, yeast, green algae, cyanobacteria, and mosses.

For years, however, he refrained from publishing for fear that his group was inadvertently contaminating the samples. One reason for concern was the dearth of spore-forming bacteria, which cocoon themselves from freezes or droughts—the kind of hardy critters you'd expect to survive in frozen soil. So each year the researchers improved their equipment—settling on a dry drill that uses no chemicals or fluids—and refined their techniques, saving only the innermost core sections that never touched a septic surface.

Gilichinsky's group started revealing its findings in the mid-1980s, including a report of viable microbes from 3-million-year-old permafrost. Surprisingly, they have found few microbes adapted exclusively to life in the cold. The ones they've dug up appear to be mostly run-of-the-mill species, which survive in thin films of water that stay liquified even a dozen degrees below zero. Other strains may live in arctic cryopegs—underground marine ponds, dozens of feet in diameter and several feet thick, sandwiched between ice-hardened permafrost. Clinging to these life rafts, the microbes may derive nourishment from the ever-so-slow leaching of minerals and gases from the sediment into the water.

The biggest puzzle is how the microbes cope with DNA damage from natural radiation in the environment. Gilichinsky believes that the microbes have sustained a modicum of unexplained metabolic activity—primarily to repair DNA and purge toxins—over the eons, although he admits he doesn't know how. While the idea is far from proven, Gilichinsky's findings show that many things scientists thought were constraints on microorganisms are simply wrong.

So far, says Gilichinsky, none of the microbes unearthed at his site have resembled pathogens. Indeed, most bacterial pathogens—with anthrax, which can kill the people it infects, an alarming exception—do not develop tough shells that would allow them to persist for millennia. He doubts that his team will unleash any scourges, ancient or otherwise, from the permafrost. And scientists in Antarctica have found no ancient pathogens in the samples obtained by drilling deep into the ice sheet. Indeed, says Sabit Abyzov of the Institute of Microbiology in Moscow, "we use masks to prevent us from contaminating the ice core."

But studies of icebound bacteria may be missing the true threat: deadly viruses.

In July 1999, a team led by Scott Rogers of the State University of New York, Syracuse, announced it had detected RNA from the tomato mosaic tobamovirus, a plant pathogen, in 140,000-year-old ice in Greenland. (His group has also isolated more than 200 kinds of fungi, some 140,000 years old, from the ice. Although scores of fungi are human pathogens, none of them turned up in the Greenland samples.) The tobamovirus RNA, more fragile than DNA, almost surely came from viral particles, although it's unclear if they could still cause an infection, says Rogers. Indeed, notes E. Imre Friedmann of Florida State University in Tallahassee, viruses, lacking an active metabolism, "should be tougher survivors than bacteria."

Rogers's study raises the possibility that the ice sheets may serve as a viral reservoir that, with global warming, could continuously release ancient forms of microbes. And if any of these strains are particularly virulent, they could be a source of new disease outbreaks—either from the thawed viruses themselves or from related viruses that manage to pick up the genes liberated from the ice.

Also posing an unknown risk is a tried-and-true killer resting in the ice. A meat locker for countless carcasses of mammoths and other Ice Age animals, the Siberian permafrost is also a tomb for political prisoners and smallpox victims from Stalin's time. During a visit to Siberia in 1990, Friedmann and a colleague came across a human corpse protruding from thawing ground. "We couldn't determine whether it was a male or a female," he recalls. A local archaeologist, looking at the old-fashioned Yakut clothing, estimated that the person had died 100 to 300 years ago. "I immediately thought of smallpox, of course," says Friedmann. But "the body was buried fast and the matter forgotten." Thousands of these victims may lie in the permafrost, waiting for global warming to set them free.

Larry Agenbroad, a scientist on Bernard Buigues's team, discounts the notion that he or his colleagues will unleash a pathogen from the ice. "Many animals defrost naturally each summer in Siberia," he says. "Where are the plagues?" Besides, he notes, previous explorers, including Japanese mammoth hunters, have dug up mammoth flesh. "I don't see anyone over there dying of mammothitis," he says.

— — —

Not a pleasure destination, Wrangel. Some 120 miles off the coast of northeast Siberia in the East Siberian Sea, it experiences perpetual darkness in midwinter, when the wind chill can plunge as low as −100 degrees Fahrenheit. It emerges for a fleeting thaw during the summer, when it receives enough feeble sunshine to nourish a few species of shrubs, grasses, and forbs.

During the Cold War, access to Wrangel, like other Siberian outposts, was restricted to military personnel and others cleared by the Soviet authorities. Even scientific trips to the islands were mainly forbidden. The military pulled up its stakes on Wrangel over a decade ago. Now only a few hardy individuals employed by the Wrangel Island State Reserve spend winters there, in a tiny settlement called Ushakovskoe, on the southern shore.

Sergey Vartanyan went to Wrangel in the summer of 1989, after receiving his *candidat nauk* degree, the Russian equivalent of a Ph.D., from the Institute of Geography in St. Petersburg, where he studied radiocarbon dating. Vartanyan was interested in gathering mammoth bones for radiocarbon analysis. At the time, the youngest known mammoth bones were from the Taimyr Peninsula and had been pegged at around 9,600 years old, several centuries after the end of the Pleistocene. Vartanyan was hoping that some of the bones on Wrangel that looked fresh might nudge that date a bit closer to the present. "All radiocarbon people like to find either the youngest or the oldest bones," he says. He collected mammoth bones from what appeared to be sediments from the modern Holocene on the slopes of hills eroded each spring. In particular, he looked in sediments near ice wedges, curtains of ice that accelerate the erosion process. Vartanyan noticed something odd right away: many of the adult mammoth bones and teeth looked small enough to have come from babies.

Vartanyan returned to St. Petersburg that September and put the bones through radiocarbon analysis. He was shocked to discover the high ratio of radioactive carbon–14 to carbon–12 in five of his samples. Four tusks and a tibia were from mammoths that lived between 4,740 and 7,380 years ago. He repeated the procedure and got the same dates, as did a lab at the Institute of Geology in Moscow. But he was reluctant to publish his findings based on so few samples. "I didn't know what to think," he says.

The next two seasons Vartanyan hauled back several hundred more pounds of mammoth bones, tusks, and teeth from Wrangel. Nearly all the bones and tusks he and his Moscow colleagues dated were from the Holocene, with the youngest only 3,730 years old. And nearly all the bones were from small mammoths, which Vartanyan and his seasoned colleagues concluded were from a race of pygmies (a notion later debunked by Dick Mol and others). This seemed to be powerful evidence that mammoths were becoming smaller as the species guttered out. And it dovetailed with findings from other islands. The Dutch paleontologist Paul Sondaar had predicted such a phenomenon from his work in Indonesia in the 1970s, hypothesizing that pygmy animals would arise in three groups: elephants, hippos, and deer. The shortened lower limb bones allow these herbivores to climb steeper slopes—island terrain—than their big-boned cousins.

Confident at last, Vartanyan and his colleagues published a report in *Nature* in 1993. In an accompanying commentary, Adrian Lister, a biologist at University College London, called the Wrangel bones "one of the most extraordinary fossil finds of recent times." Scientists on all sides of the extinction debate hailed the findings. Overchill proponents viewed Wrangel Island as a holdover of the Pleistocene: some argued that patches of its vegetation resembled mammoth steppe, so it was no surprise that mammoths were able to persist there for thousands of years after the species died out elsewhere. Successive generations of Wrangel mammoths, they say, attained a shorter stature as their food supply gradually diminished. Overkill advocates rejoiced because the close dates between the last mammoth and first human presence fit their hypothesis nicely. Just as the first New Zealanders appear to have feasted on the flightless moa until the bird was wiped out and the early settlers of Cyprus killed off the pygmy hippo, the hunters who first reached Wrangel could have finished off the mammoths.

MacPhee thought Wrangel was a good place to test for hyperdisease. And he knew that Vartanyan was just the person to collect samples on Wrangel. However, MacPhee had a hard time funding the trip until a wealthy British businessman whom he had met a few years earlier, on a train trip across China sponsored by the museum, came through with some seed money. The U.S. National Science Foundation then followed with a grant for $50,000 to cover the remainder of the cost—with helicopter rates running $5,000 an hour in that lonely corner of Siberia, MacPhee figured he had just enough to get a five-person team there and back.

In August 1998 they flew from Moscow to Mys Shmidta, an outpost on the Chukotka coast. There they chartered a helicopter for the 250-kilometer ride to Ushakovskoe and drove to a campsite in the north of the island. "It's really the edge of the earth," says MacPhee. Their home for the next eighteen days would be Puma Three, an 8- by 17-foot cabin named after its radio handle. Wrangel has no trees, yet a sign outside the cabin jokes: Wrangel Island: Eastern Forestry Division. In their refrigerator—a pit dug in the permafrost—they stowed their main course for every meal, salmon and reindeer meat.

During the 20-hour midsummer days, the five researchers —Vartanyan, MacPhee, Clare Flemming, Alexei Tikhonov of the Institute of Zoology in St. Petersburg, and Jeff Saunders of the Illinois State Museum—spread out on foot across the tundra, hauling back to Puma Three promising bones, teeth, and tusks. "You're completely at the mercy of the weather," MacPhee says. "Only two of the days could we take off our parkas; the rest of the time was rain, sleet, and snow." Each day's walk was a slog through a patchwork of knobby tussocks and puddles in the melted permafrost. "It would be comparable to walking across a plowed field in the rain, with gale-force winds blowing at your back," says MacPhee, "while carrying 100 pounds of mammoth goodies." To cover more ground they set up camp on the Shumanaya River, 7 miles southeast of the cabin, spending four fitful nights in tents and wondering if the island's polar bears would pay a visit.

Bad weather postponed their scheduled pickup one day, then another day, then another. "Since the civil aviation apparatus [in Russia] has basically fallen apart, the town wouldn't risk its one helicopter unless the conditions were good," says MacPhee. The team grew nervous as the date approached for their charter flight from Mys Schmidta back to Moscow. The morning of the flight, the airport in Mys Schmidta finally gave the okay for the helicopter to pick up the team and held up the charter flight for three hours for the team's return. "Just imagine in North America, you have some idiots on an adventure off in some place. You're going to wait on the ground for these guys to show up? I don't think so." They were expecting nasty looks from the other passengers when they finally boarded the plane. Instead, says MacPhee, "there was just cheering, passing the vodka around."

They were lucky to get out when they did. Originally, the researchers had planned to share expenses with a New

Zealand film crew doing a documentary on polar bears, but the TV people weren't ready to go in August. The film crew made it out to Wrangel in mid-September, a few weeks after the researchers left. No sooner had they arrived than winter did too—early. A sharp cold snap brought strong winds and snow, stranding the crew in their tents for two months. With their food supplies nearly exhausted and the helicopter in Mys Shmidta grounded, the men became desperate. Says MacPhee, "They were ringing up all their embassies by satellite phone, telling them, 'Get us out of here!'" News of their plight was broadcast around the world, prompting the Russian Air Force to launch an expensive rescue. Ironically, he says, "they didn't even get out to film."

The researchers hauled a few crates of mammoth bones back to St. Petersburg, including an ulna so fresh it exuded greasy marrow. MacPhee took samples of bone to take back to the United States for radiocarbon dating and possibly hyperdisease analysis.

— — —

Taking on the task of analyzing the Wrangel bones was MacPhee's resident DNA expert, a young molecular biologist named Alex Greenwood who had just joined the lab. Greenwood had trained under a noted molecular biologist, Svante Pääbo, whose group had published a stunning paper in 1997 in which they described a sequence of DNA from bone of a Neanderthal man. The small snatch of DNA—extracted from mitochondria, the cell's biochemical powerhouses—was very different from the corresponding stretch of DNA in modern humans, adding to the hunch that these Stone Age humans were more distant relatives of ours—perhaps even a different species. Notwithstanding that remarkable bit of molecular legerdemain, no complete strand of ancient mitochondrial DNA had been extracted until early in 2001, when Alan Cooper of the University of Oxford and his colleagues announced that they had deciphered the mitochondrial DNA from two kinds of moa, flightless birds that disappeared from New Zealand around 500 years ago, the apparent victim of zealous human hunters.

For every ancient DNA success story, however, there's a high-profile failure or embarrassment. The field is replete with findings gone bust, often because samples were contaminated inadvertently with modern DNA. "One's reputation hangs on a thin thread, and one's colleagues are unforgiving," MacPhee says. Particularly suspect are claims of DNA having been extracted from insects in amber (fossilized tree resin) or from dinosaur bones tens of millions of years old. DNA decays with time under a relentless assault from acids and corrosive forms of oxygen that attack the base pairs forming the long DNA chain, corrupting them. Even in the permafrost, where such degradation is slowed, researchers calculate that any DNA older than 100,000 years should have been marred beyond recognition by today's standard tool for detecting DNA fragments, the polymerase chain reaction (PCR). In this procedure, a probe is used to latch onto a specific DNA sequence, which is then copied millions of times by an enzyme, allowing it to be analyzed. The problems creep in when the PCR probe grabs modern DNA contaminating a sample—even the tiniest snippet can spoil an experiment or, even worse, mislead researchers into believing they have extracted ancient DNA. Greenwood and MacPhee knew that they had to proceed cautiously. "Mammoth DNA is in such crappy condition that it easily could be swamped by contaminants," says Greenwood. "That's what makes the whole process really, really scary."

Most ancient DNA specialists have used brute force to

overcome this impediment. Rather than home in on the cell's nucleus, which contains billions of base pairs, they have instead focused on the much shorter sequences—16,000 base pairs or so—inside the cell's mitochondria. That would seem nonsensical—if you want to score a hit, why aim your pistol at the weathervane when you can easily nail the barn? The answer is that there's only a single double-stranded copy of nuclear DNA in each cell, and this DNA is composed of tens of thousands of individual genes; any PCR primer would have to focus on a specific stretch of this miles-long sequence of base pairs. Mitochondrial DNA, on the other hand, comprises only a handful of genes, and there can be thousands of mitochondria in a cell containing thousands of copies of these genes. Consequently, there's a much higher chance that even in degraded ancient DNA samples, at least some of these genes have survived the long march of time. Mitochondrial DNA has an added advantage: determining the sequence of base pairs in the same mitochondrial gene from many species has allowed researchers to assemble a molecular maternal family tree of extinct and modern life forms. This tree complements (or, more often, conflicts with) the family tree developed after comparing the shapes of bones of extinct and living animals.

But nuclear DNA contains the vast majority of genes that give an organism its appearance and influence how it behaves. Everything from hair color to our own species' potential for mathematical or musical genius is dictated by genes in the nucleus. "It was thought to be almost impossible" to retrieve genes from the nucleus of samples so old, Greenwood says, as the odds were that any particular gene targeted by a PCR primer would have degraded and likely have been undetectable. Nevertheless, Greenwood wanted to blaze a new trail. Working with his mentor, Pääbo, and two other scientists, he set out to be the first person ever to extract fragile nuclear DNA from mammoth cells.

Greenwood began with the molars of two mammoth skeletons excavated from permafrost near Engineer Creek, Alaska, as well as a jawbone from mammoth remains found on the Arctic Ocean's Novosibirskie Islands, off the coast of central Siberia. The first step was to blast his workroom with high levels of ultraviolet radiation to break down any potential contamination, including human DNA from sloughed-off skin cells or oils left by touching the lab benches with bare hands—the same oils that fingerprint experts dust for. Next, he ground up fragments of the samples, then removed the minerals by dissolving the powder in a solution containing EDTA, a chemical that grabs any mineral it comes into contact with and precipitates out of solution. After flushing out the minerals, Greenwood added an enzyme, proteinase K, which set about dismembering the proteins in the bone powder.

He was then confronted by that gremlin of ancient DNA: the peculiar chemistry that goes on inside cells after death. "Tissue decay creates a lot of compounds that wouldn't normally be there in living tissue," Greenwood says. Particularly vexing are compounds formed when blood decomposes. Often, when using a chemical called phenol to extract DNA from dead tissue, samples take on a purplish or brownish tinge. "That's a bad sign," Greenwood says. It usually means that the sample is spoiled and the odds of retrieving DNA are minuscule. Occasionally, however, the sample was clear, and he could proceed to the next step: adding silica and a chemical, guanadinium isothiocyanate, that forms a salt bridge between the DNA and the silica. Everything else could then be rinsed away with alcohol. Like hosing down a sandcastle, all it takes is a little water to shatter the salt bridges.

Greenwood used PCR primers designed to home in on unique sequences in two genes: the 28S ribosomal DNA gene, of which a few hundred copies exist in vertebrate genomes, and the von Willebrand factor gene, one pair of which is found in each cell. (Defective copies of the von Willebrand gene lead to the most common form of inherited hemophilia.) To his delight, Greenwood succeeded in recovering fragments of both genes from mammoth DNA. Comparing their sequences to those from the same genes in African and Asian elephants, he added new evidence to previous reports that the mammoth was a closer relative of Asian elephants.

While MacPhee's Wrangel samples have not yet surrendered a potential hyperdisease pathogen, they have yielded other intriguing viral DNA: so-called endogenous retrovirus-like elements, viral sequences that worm their way into the genomes of higher organisms but that appear to be benign. Before Greenwood, it appears, no scientist had bothered to look for these elements in fossils. For this experiment he tapped the freshest-looking Wrangel bone—the greasy tibia from a 4,590-year-old mammoth—as well as a molar from the 13,780-year-old Engineer Creek, Alaska, mammoth. Greenwood found that some retroviral elements in his mammoth samples were identical or clearly related to elements in modern elephants. This was surprising, considering that endogenous retrovirus-like elements have no known advantage to an organism and should be routinely flushed out of the genome. Yet somehow, these particular elements had persisted for millennia in different species across the globe.

More important for Ross MacPhee's quest to find a hyperdisease pathogen, finding the DNA-based retrovirus-like elements in ancient DNA was powerful evidence that pathogenic viruses could, in principle, be found in mammoths that died during the species' last stand. That only exacerbated MacPhee's frustration: the technology was there. If his mammoth killer existed, he could find it—if only he could get back to Siberia.

THE BIG LIFT

Larry Agenbroad got a phone call one day in June 1974 that would change his life. A few days earlier, a fellow named George Hanson was bulldozing land for a housing development on the outskirts of Hot Springs, in the Black Hills of southwestern South Dakota, when he unearthed some large bones. Hanson's son stubbed his toe on one, which he recognized as a mammoth tooth. Hanson tracked down the nearest mammoth expert, who happened to be Agenbroad, then a geology professor at Northern Arizona University. When Agenbroad arrived at the site early the next month, he was intrigued to discover that the bones belonged to at least four mammoths.

The following summer the landowner, Phil Anderson, allowed Agenbroad and a group of volunteers to excavate further. One member of the team discovered a mammoth skull preserved with the tusks intact. They were on to something big, Agenbroad thought. Convinced that the site should be set aside for science, with Anderson's blessing, Agenbroad helped the citizens of Hot Springs set up a nonprofit corporation and raise more than $1 million to buy the land and erect a shelter over the nascent dig. "It's pretty scary to think about what the site might have been if those first bones were never found," Agenbroad says. Of the nearly 1,500 locations in the Americas where mammoth bones have been found, none is as impressive as the Mammoth Site at Hot Springs. From the beginning, Agenbroad has served as the site's director, overseeing an ongoing excavation at a former sinkhole that so far has yielded fifty-two mammoth skeletons.

Dick Mol had spent two summers in the early 1990s as a visiting scientist at the Mammoth Site and knew Agenbroad well. So when the Discovery Channel—which was planning to film the excavation of the mammoth remains on the Taimyr Peninsula—told Mol and Bernard Buigues that they wanted an American on the team to entice a U.S. audience, Mol knew just whom to ask. He had seen how the volunteers at the Mammoth Site adored the charismatic geologist. Agenbroad could serve as an erudite and enthusiastic spokesperson in the United States, Mol thought.

In March 1999 the documentary's executive producer, Maurice Paleau, flew to Agenbroad's winter haven—Flagstaff, the home of Northern Arizona University—to size him up for the documentary. "We hit it off really well," says Agenbroad, who keeps two mammoth femurs in his office. The femurs make an odd pair: one, from a gigantic Columbian mammoth, is more than a yard tall; the other, from a pony-size pygmy mammoth, less than 2 feet long. He handed the director one of his business cards, which features a cartoon of a mammoth with the words "Have Mammoth??? Will Travel!!!!" Agenbroad was just the character they were looking for, so Paleau proposed that he meet with Buigues at the International Mammoth Conference that May in Rotterdam.

Mol, meanwhile, took a break from organizing the mammoth convocation to fly to Khatanga in April to see at first hand the upper and lower jaws and the teeth that Buigues had dug up in his preliminary excavation the previous spring. The radiocarbon dates of the samples Buigues had given him in the fall showed that the mammoth had died around 20,380 years ago. The hefty size of the tusks, which Mol had examined in Buigues's photographs, clearly indicated that the mammoth was a male weighing around 15,000 pounds. Any further details, Mol knew, would only emerge from direct examination of the fossils.

When Mol got to Khatanga, Buigues showed him the tusks. "They were beautiful," he says, "the best preserved I had ever seen." Mol also examined the jawbone, complete with the sixth set of molars—mammoths had a maximum of six sets—so he knew it was an adult. He could get a more exact determination of the animal's age at death by measuring the amount of wear and plugging those numbers into a formula based on the wear of African elephant teeth at known ages. He found that the mammoth died in its prime, when it was roughly forty-seven years old.

Convention dictated that Buigues, as discoverer, name the mammoth after the place where it was found. Taimyr and Khatanga were assigned to previous finds, so one possibility was to name it after the nearby Bolshaya Balakhnya River. That sounded unappealing to Buigues, who already felt a bit uncomfortable having to name the find. While he had the rights to excavate the mammoth—he had obtained formal permission from the regional administration to mount the dig on state land, with an explicit agreement that the excavated remains would not leave the Taimyr Peninsula—he wasn't its discoverer. But conferring a geographic name was not a hard-and-fast rule. For instance, the Adams mammoth was named after its discoverer. And in 1988, after retrieving a baby mammoth carcass that a ship captain had spotted on the bank of the Yuribeteyakha River on northern Siberia's Yamal Peninsula, Alexei Tikhonov named the desiccated female calf —not nearly as well preserved as the famous *mamontyonok* Dima—after his daughter, Masha. Helping Buigues, Mol had no trouble hitting on a name for the Taimyr find that would please everybody: Jarkov, after the family that had led Buigues to the site.

Mol returned home after six weeks in Khatanga, knowing that he faced a difficult selling job: persuading his colleagues at the mammoth conference later that month that the upcoming excavation was not a publicity stunt. He had already received several inquisitive, but skeptical, e-mails. "Everyone had heard about this project, but they only heard about a crazy Frenchman who had a lot of money and who wanted to extract a mammoth," Mol says. He remembers the reaction he got from Andrei Sher when he picked him up at the Amsterdam airport. Sher said, "I understand from the program that we are going to hear about the mystery mammoth excavation," Mol recalls. "This was a sign to me that nobody took it seriously."

The conference, held at Rotterdam's Natuurmuseum, drew more than a hundred mammoth experts from around the world. Presentations ranged from genetic analyses of mammoth DNA to possible migration routes of woolly mammoths as they fanned out across the globe. Studies also looked at the mammoth's contemporaries, including an imaginative presentation on the fighting behavior of woolly rhinos based on observing their modern counterparts. Most thrilling, however, was the announcement of the Jarkov discovery, accompanied by a video clip of Buigues thawing out the skull.

Sher says the presentation helped alleviate some of his

concerns about the project. An important new piece of information that surfaced at the meeting was the dig's location near the Bolshaya Balakhnya River. Not covered by glaciers during the Great Ice Age—thus presumably primarily mammoth steppe—the Bolshaya Balakhnya region is a well-known graveyard for mammoths, musk oxen, Pleistocene horses, and other extinct creatures, suggesting the find could be promising. And the scientists were particularly impressed by the plan for airlifting the specimen to a place where it could be thawed slowly to extract the most scientific information possible. That contrasted with the usual practice of using a hot water hose to melt the frozen ground around a mammoth, thereby destroying any clues in the surrounding sediments and possibly damaging the specimen as well.

Many experts were reassured to learn that the expedition's chief scientist was the paleoanthropologist Yves Coppens, who codiscovered the Lucy hominid bones and was an expert on Probosicideans. Also encouraging was that the project's aim was not to clone a mammoth, as some newspaper articles had claimed, but to understand how the mammoth lived and why it went extinct. Of course, the animal's DNA was a highly sought-after prize. While Buigues's team was ready to share the freshest samples with scientists not affiliated with the expedition who wanted to try cloning the mammoth, the team would use the DNA to probe further the evolutionary relationship of the woolly mammoth to elephants and other relatives like the manatee. And if Buigues could find carcasses of mammoths that died at the end of the Great Ice Age, their frozen DNA might help solve the extinction mystery. Overchill proponents would get a boost if the genetic diversity of the creatures was declining in lockstep with changes in vegetation, while hyperdisease enthusiasts would thrill to find the genetic traces of an ancient pathogen.

But the DNA jocks wouldn't have a monopoly on the projects with the sexiest implications. More mundane analyses, such as studying the pollen and other plant material trapped in Jarkov's hair and in the surrounding sediments, could offer insights into the Ice Age climate in that part of the Taimyr Peninsula. Some grasses would more likely have grown on a cold, dry mammoth steppe than on a land that got more precipitation. Along with an analysis of the plant remains in Jarkov's stomach—the animal's last meal—this could provide a crucial baseline for comparing the plant life on Taimyr before and after the Great Ice Age. Correlating the plant turnover from one Siberian region to the next with declines in mammoth populations or with the apparent health of individual mammoths when they died could make or break the overchill hypothesis.

To Agenbroad, one of the more exciting aspects of the project was being able to bring together scientists from different disciplines to figure out how and why the Jarkov mammoth died, and what the circumstances of its death might have meant for the species. Working in one branch of science is like developing film in a darkroom and seeing the vaguest outlines of an image, he says. "Unless you bring in the other disciplines, you may never see the whole picture."

To give the expedition more international panache—and score points with Russian officials and the general public—Mol and Buigues decided to invite a young Russian to join the team. Mol, who knew the mammoth scientists at the Museum of Zoology in St. Petersburg, proposed Alexei Tikhonov, the protégé of the venerable Nikolai Vereshchagin. In the meantime, Buigues missed the chance to meet Larry Agenbroad, who was laid up by blood clots in his legs and couldn't go to the mammoth conference. So in June, Buigues, Mol, and the film crew flew out to Hot Springs a few days after Agenbroad had arrived for the summer field season. Buigues noticed how well Larry and his wife, Wanda, worked together at the Mammoth Site. Out of the blue, he asked Wanda if she, too, would like to come to Taimyr. "Bernard blew Wanda away by inviting her," says the gravelly voiced geologist. "She was the only petunia in our onion patch out there."

Fast approaching their silver wedding anniversary, Larry and Wanda Agenbroad, a retired teacher who is now a research associate at the mammoth site, still flirt and kid each other like newlyweds. But Larry is quick to point out that the high school sweethearts encountered some bumps along the way, including a melodramatic letter that Larry remembers getting when he was in the navy stationed in Morocco in the early 1950s. "She said she was going to forget me and go and be a stewardess," Larry says. He thought they were finished, but when he came off duty three years later, he says, "I called her from New York and she said, 'I bought the bridesmaid dresses today.' I said to myself, 'Uh-oh.' I was going to go back to Morocco if I could have made a living there. You can have four wives in Morocco," he says, teasing Wanda, who smiles back.

After his military service, Agenbroad got a master's degree in geological engineering from the University of Arizona, then took a job in the oil industry, where he used mathematical modeling to prospect for undiscovered reserves around the world. "It was mind-deadening work," he confesses, although he enjoyed the travel. Eventually he returned to the university, earning a second master's in archaeology and a doctorate in geology. He began studying the prehistoric hunters in North America who stalked mammoths. His passion veered toward the beasts themselves—deciphering their life histories and evolution from the fossil record and reconstructing the environment in which they lived.

When the Agenbroads arrived in Moscow on September 16, 1999, Larry was disturbed to find that the weather was warm enough for shorts. If it was anywhere near that warm in Siberia, the team would not be able to excavate the block of permafrost containing the mammoth and keep it frozen. The next evening, they were whisked to a military airport for the flight to Khatanga. For the seven-hour, 2,000-mile trip, Buigues had chartered an Ilyushin plane that, they were told, served in the 1950s as a strategic bomber capable of reaching the United States. Arriving at the airport, they found the plane packed with Russians who had cut private deals with the plane's crew. (A similar experience to Goto's in Khabarovsk two years earlier.) The interlopers and their baggage left almost no room for the expedition, a team that now included the production staff of the Discovery Channel, which had bought world broadcasting rights to the documentary. It was the only time Agenbroad recalls seeing Buigues visibly angry. "I remember him saying, 'What is this crap? This plane's not leaving until I get my team and my baggage up there!'" The negotiations ended fruitlessly, and the team crammed into the small section of the plane reserved for the crew. "We had baggage stacked around us; no seat belts, nothing," says Agenbroad. "We would have never gotten off the ground in the United States."

The team rested after the eastbound red-eye, then dined with the administrator of the Khatanga region in the town's restaurant, which "is only open when they feel like it," Agenbroad says. Dinner that night and every night thereafter consisted of various combinations of reindeer, fish, and potatoes. But the uninspiring fare didn't dampen Agenbroad's spirits. He recalls his elation after being told that the weather forecast that night called for freezing temperatures. "Everybody cheered," he says. "We could dig."

Two days later they took a helicopter one and a half hours northwest to a spot on the tundra exactly 477 miles north of the Arctic Circle. But when Agenbroad saw the Russian laborers digging around the frozen sediments containing the find, his heart sank. Muddy water was pooling in the shallow trenches, and he was afraid that the water would accelerate the thawing and expose the mammoth. But he grew excited when he spotted long black hairs sticking from the top of the block. When he stroked the beast's hair, he confessed later, "it was like holding your child for the first time."

The team saw the season's first snowfall that night as a good omen. But the digging went slowly because a generator Buigues had brought wasn't powerful enough to run the two compressed-air jackhammers. So the workers, in temperatures that had dropped to around minus 25 degrees Celsius, used pickaxes and shovels to chip away at the concrete-like permafrost. Agenbroad spent a week at the site and a week in Khatanga reading, eating, and talking with Mol and Buigues. Agenbroad warmed to the awestruck children who had never seen an American in the flesh. One day, he says, a young Dolgan girl, maybe nine or ten years old, was trailing a few steps behind him. "I turned around and she said, 'My name is Anna.'" Her English vocabulary exhausted, she approached Agenbroad a week later with a new line: "Your name is Larry." Smitten, he surrendered a coveted package of Lifesavers.

One afternoon in Khatanga's Musk Ox Research Center, a Dolgan woman sang in lilting melodies that reminded him of Native American music. Some of the songs were sad, and while Agenbroad couldn't understand the words, he grasped their sincerity. Many Dolgan who choose to live in Khatanga instead of on the tundra are stuck in dilapidated wooden shacks. Others who have resumed their nomadic lifestyle depend on domesticated reindeer for survival. They fiercely guard their reindeer, which pull sleighs as well as their homes, or *baloki,* in which they winter on the tundra. (Buigues has had a *balok,* reindeer skins stretched over a cubic wooden frame, built for himself.)

You don't have to sled far outside Khatanga to see wild reindeer. "They're on every horizon," says Agenbroad. One day he accompanied Gennady Jarkov on a reindeer hunt. After the Dolgan shot one and dressed it, he let Larry drive his three-reindeer sled back to town. Jarkov handed him what Agenbroad calls "the reindeer accelerator": a 20-foot-long pole with an ivory knob at the end. "If they aren't going fast enough, they get poked right in a strategic spot with the knob and they tuck and run," Agenbroad says. "You're going along and all of a sudden—jerk! I got rolled off the sleigh on one of those real fast starts." As Agenbroad was adapting to Khatangan life, Buigues had his hands full with logistics. A major headache was finding a helicopter to lift a permafrost block estimated then to weigh 16 tons and haul it the 200 miles back to Khatanga. The operation would require the biggest civilian helicopter in the world. Forking over piles of rubles to Aeroflot was unavoidable, but Buigues had to provide an extra favor to obtain the necessary authorization. The deal was as follows: he could raise the mammoth with the helicopter if he first brought the vice speaker of Russia's lower house of parliament, Artur Chilingarov, and a group of his friends by helicopter to the North Pole on September 25, the legislator's sixtieth birthday. It turned out that Chilingarov had led several scientific expeditions in the Arctic and was once chief of one of Russia's Antarctic bases. "You have to understand the Russian way of doing things," says Agenbroad. "We couldn't have done the lift otherwise."

There were a few extra seats on the trip, allowing Agenbroad and Mol to tag along. Agenbroad was not impressed.

Gray clouds filled the sky, and the ice over the Arctic Ocean was too thin for the helicopter to land safely. It hovered a few feet above the Pole while Chilingarov and some others hopped out for a vodka toast. The highlight for Agenbroad was the trip back to Khatanga. From the helicopter, he watched the last sunset of the millennium at the North Pole. The sun wouldn't rise again until March.

Bidding the sun good-bye reminded them that the time was growing short: the brutal High Arctic winter was fast approaching, and there would be no lift unless Buigues could speed up the digging. On Agenbroad's last day in the field camp, September 30, Buigues finally managed to procure a generator strong enough to power the two jackhammers. "They did more work in one afternoon than they were [previously] able to do in two weeks," says Agenbroad, who had to return to his teaching.

In less than a week, the excavators finished digging the trenches and started to chisel space under the block to insert steel rods for a metal frame to support the block during its flight. But Buigues had yet another difficult problem: he had not stockpiled enough fuel for the helicopter, and Khatanga was in short supply that autumn. A barge carrying the town's winter fuel had failed to make it into port before the Khatanga River froze, so it was stuck downstream, waiting for the arrival of a nuclear icebreaker to open a channel. Buigues appealed to the Discovery Channel, which came through with several hundred thousand dollars to pay for fuel to be flown from the Arctic mining city of Norilsk, a few hundred miles southwest of Khatanga.

When Buigues left for Norilsk to sort out this complicated purchase, he left Mol in charge of the excavation. Although a camaraderie had developed between the film crew and the laborers, the Russians chaffed under Mol. When

Buigues was there, people would sit around the table in the evening after supper, drinking vodka and smoking cigarettes, enjoying themselves despite the language barrier. But Mol neither drank nor smoked, and although loquacious, he came off as aloof. "I wasn't fitting in," he says. More alarming, he sensed that the Russians were growing reluctant to follow his orders. He reconciled himself to a radical solution: he took up smoking—something he had never done before. Gambling with his health paid off; the Russians respected his new addiction.

As the team waited for Buigues, the weather began to deteriorate, intensifying into a blizzard. The tent shuddered in the gusts; expedition members feared that it would be swept away, exposing them to the lethal wind chill. "We didn't know when the huge helicopter would arrive," says Mol. "We were waiting and waiting." Every morning the trenches would fill up with blown snow that had to be shoveled out—until somebody thought to cover the block with a tarp. During this tense period, Mol collected from inside the trench several fragments of aquatic plant material, including some fragments two or three inches long that had retained a greenish luster. The plants suggested that 20,000 years ago, the site was the bank of a pond or a stream. One hypothesis for the mammoth's death, says Agenbroad, is that "the guy came to get a drink and got mired."

But the TV crew was not satisfied by pond scum. When Buigues returned from Norilsk, his filmmaker Jean-Charles Deniau asked him to defrost a patch of the block to reveal a glimpse of the mammoth. Mol, not wanting to damage Jarkov more than it had already sustained, says he was against this idea, but Buigues acquiesced and turned to his trusty blow dryer. Working around the clock for three days, he and the others thawed the top few inches of a three-foot-wide section of the block, thereby liberating long bristly guard hairs, a silky layer of grayish-yellow wool, and the tawny hair of the undercoat. Mol stroked the hair. "It was like touching a living mammoth," he says. And it smelled like one too: the hairs exuded a pungent odor, like dung. "From what I could tell, the hair was still attached to skin," Mol says.

The film crew also wanted to use metal collars to attach the tusks to the permafrost block before it was raised from the pit. A betusked block gliding through the air, they argued rightly, would look much better on TV than a naked block. Buigues agreed, although later, when criticized for this bit of showmanship, he said he had attached the tusks to honor the mammoth. "I did not feel it would be appropriate for him to travel to his new home without them," he says.

The next day was Sunday, October 17—"a date I will never, ever forget," says Mol. The storm had blown itself out the night before, and the day broke sunny and calm. About ten in the morning, Mol got a call from Khatanga: the helicopter was fueled and ready to fly; he should see to breaking down the camp.

Three hours later, the MI–26 arrived. "It was an incredible machine, about 30 meters long, the width of the blades about 28 meters, making a lot of noise," Mol says. The downdraft blew up snow and made visibility on the ground difficult. Christian de Marliave, who had designed the metal sling to hold the block, grabbed a hook on the end of a cable lowered from the belly of the copter. With 100,000 kilograms of metal hovering over his head, de Marliave fastened the cable to the block and signaled to the crew. Everybody was scared.

The massive helicopter strained against the load, which clearly was heavier than the estimated 16 tons. "It seemed impossible to lift," says Buigues, who was standing on the ground, the wind from the helicopter blades giving his fingertips frostbite. For several minutes the helicopter strained, but the block hardly budged. Mol imagined that at any second the helicopter would come crashing down. From his vantage point in the cockpit, the Discovery Channel's production assistant Dirk Hoogstra was also terrified. "More than once it felt as if our load was going to pull us down to Earth. Each moment seemed to last forever," he reported afterward on *Discovery Online.* Even a minor accident that didn't harm anyone might have had severe repercussions. "People would say, 'Look at that adventurer and that amateur doing crazy things in Taimyr,'" Mol says. The outcry would have killed the prospects for other Western paleontologists hoping to do field work in Russia.

After nearly ten minutes of a tug-of-war, the block began to lurch upward before finally rising into a clear sky tinged with purple twilight. Mol gasped; he saw the block's iron frame, designed to support 30 tons, start to bend. But then, just as suddenly, the frame stiffened—and held all the way to Khatanga. Mol and the rest of the team from the base camp rode in a smaller helicopter that preceded the MI–26 back to the airfield. A few minutes after they arrived, the block appeared on the horizon, the tusks still attached. The pilots later revealed that the block weighed nearly 23 tons—3 tons more than the helicopter's official lifting capacity.

"When we set the mammoth on the runway in Khatanga," Mol says, "I knew we had written history." Two hundred years after the woolly mammoth was described as an extinct creature and the Adams mammoth began to emerge from its frozen crypt on the Laptev Sea, Bernard Buigues had hauled frozen Ice Age remains for the first time to a safe place where they could be thawed gradually from their icy tomb. "This was a milestone in paleontology," Mol claims. "People will speak of this discovery 100 years, 200 years in the future."

— — —

The Dolgans, however, weren't celebrating. None of them had participated in the excavation for fear of its bringing bad luck. "They told Bernard to sacrifice a white reindeer," says Vladimir Eisner, an associate of Buigues's in Khatanga. Not only would such a sacrifice ward off misfortune, the Dolgans said, but it would also entitle Buigues to the bones. Buigues balked at killing a reindeer of any color, and he refused to consider the alternative offered by the Dolgans: sacrificing a dog. Indeed, one of his team members had already lost a dog to a Taimyr blizzard. They finally suggested that Buigues throw some coins in the hole. That he agreed to, but in retrospect the Dolgans viewed the gesture as a failure. In early 2000, several people met with accidental deaths in the Dolgan village of Novorybnoye, a favorite stop for Buigues to buy bones and tusks and get information on promising mammoth sites. The villagers blamed the deaths on him and asked him to stop digging. "They told him, 'Don't go back to the tundra. People die,'" says Eisner. "This is a man the Dolgans had known six or seven years. Buigues would give them his own coat if they needed it. That's how strongly they felt about disturbing the remains."

The Dolgans weren't the only people voicing doubts. Not long after the triumphant conclusion of the $2 million expedition, one of the team's own scientists claimed that they were unlikely to find much of a mammoth in the block. Tikhonov, who had spent a week at the site in October 1999, claimed that only the top two feet of the excavated block consisted of the sandy earth that was likely to preserve tissues. The remainder of the block could only be an ice wedge, he asserted, a crack that widened slowly as it filled with snow and water. No carcass could have been preserved in such a formation. Moreover, he pointed out, the bones found at the site before the lift were clean, with no signs of attached muscle. "All of this suggests that the find is not very important, not interesting for science," he said at the time.

Others disagreed. Agenbroad says that after Tikhonov left Taimyr, further digging uncovered a layer of stream gravel, showing that the bottom half of the block was not an ice wedge. One Western team member suspected that the Russians were purposely playing down the Jarkov mammoth's significance: "I'd have been a little bit miffed if we had a Russian expedition come over to our country and work in my backyard when we couldn't afford to." Only thawing the block would determine whether Jarkov was a find worth squabbling over.

THE DNA MENAGERIE

Unlike Dick Mol and other members of the Buigues expedition, Larry Agenbroad has staunchly defended the idea of cloning a mammoth. He sees no harm in keeping a twenty-first-century mammoth in an artificial environment as long as it's cared for. "Too many people saw *Jurassic Park* too many times. This is not a carnivorous animal that's going to be cloned," he says. "We're talking about a hairy baby elephant. Unless you're a vegetable, you have nothing to worry about."

The birth of Dolly the sheep, the first mammal cloned from an adult cell, ushered in the age of cloning in 1997—along with the possibility of resurrecting extinct species, including the mammoth. "The idea is not so far-fetched when you consider the technology we have today and will have in the future," says John Critser of Indiana University, who notes that all along, "the leap of faith is finding the viable sperm or eggs." Others are much more skeptical. According to one prominent reproductive biologist, "Too many of these scientists are trying to fly before they can crawl."

Indeed, even with good mammoth DNA in hand, Goto or anyone else embarking on a cloning project would face formidable hurdles. The experiences with Dolly and the cloned animals that have followed in her hoofsteps demonstrate that cloning is a challenge even for well-studied species like sheep, cows, and mice. Cloned embryos suffer a high abortion rate, while those brought to term are more likely to die soon after birth from defects in the heart, lungs, and other organs. It's still unclear what factors are responsible for these failures, although there's a growing sense that a process called genetic imprinting is critical. Inside a fertilized egg, molecules somehow designate which genes from the father and which from the mother will be turned on. Without DNA from both parents, this process gets derailed. Imbalances in imprinting can cause the placenta to form poorly, which can harm a developing fetus.

However, there are encouraging developments as well, including the first success at implanting the embryo of a cloned endangered species in another species. In early 2000, the embryo of a gaur, a wild ox native to Southeast Asia, was successfully implanted in a cow. A day after its birth in January 2001, the baby, named Noah, came down with scours, a diarrhea common in newborn calves. The vets diagnosed a bacterial infection and treated Noah with antibiotics, but he died a day later. Nevertheless, the experiment, undertaken by the biotechnology firm Advanced Cell Technology in Worcester, Massachusetts, was hailed as a success in showing the promise for saving endangered species and possibly even resurrecting extinct animals.

It would be wise to clone certain endangered species before they reach the point of no return. "Contrary to the most common version of the Noah's Ark legend, a single pair is usually not adequate to found a viable population," writes the conservation biologist Thomas J. Foose in *Riders of the Last Ark.* Indeed, more ancient accounts of the flood suggest that Noah took aboard not just one but seven pairs of each species. If so, Noah knew his population biology: on average, seven breeding pairs should harbor more than 90 percent of the genetic diversity of the population they are drawn from. Clones resurrected from frozen DNA samples could broaden the gene pool by bringing lost genes back into the population, thus providing the range of genes necessary to replenish a species that is down to its last few individuals.

The giant panda is poised to become the first species cloned for the sole purpose of increasing its genetic diversity. Even though major swaths of the panda's habitat are now protected for the thousand or so individuals left in the wild, the Chinese government is scrambling for some insurance. As part of a program to develop in vitro fertilization techniques for the panda, it has pledged to clone the species by 2004. The geneticist leading the effort, Chen Dayuan of the Chinese Academy of Sciences, has claimed he is ahead of schedule. In June 1999 his team announced a milestone: they fused skeletal muscle and mammary gland cells from a dead female panda with the egg cells of a Japanese rabbit stripped of their nuclei to produce blastocysts, which are early embryos.

The Chinese group is working with Advanced Cell Technology to try to repeat this feat with egg cells from American black bears. If they get an embryo, it would be implanted in a surrogate mother bear. Such an approach may also pay off for zoos, which have notoriously poor records for breeding pandas. Cloning ten copies of a single stud panda, then mating the clones with ten different pandas, would preserve 95 percent of the cloned panda's genes.

The panda project stands a good chance to succeed because it builds on years of research on the animal's reproductive physiology and flaws. Scientists have learned, for instance, about an endocrine disorder in females that can impede fertilization, and about a common problem in males, shrunken testicles. They also have a fair handle on the biology of black bears, which would carry the cloned panda fetuses to term. Indeed, in such an experiment it's more important to know the reproductive foibles of the surrogate than the biology of the clone; after all, the clone's development is moot if researchers cannot judge when and where to implant the clone's egg. That's a problem, because the reproductive system is a mystery in many creatures. Thus more exotic endangered species that might be good targets for cloning, like the Sumatran rhino or the river dolphin, may have to wait until the reproductive biology of suitable surrogates is worked out.

Cloning living animals stands a higher chance of succeeding than cloning dead ones, if for no other reason than that the DNA from fresh cells, if properly handled, remains whole and arrayed in chromosomes, whereas the DNA from dead cells has generally broken up. Even in the Siberian permafrost, DNA molecules lie shattered, like a windowpane dropped from the top of the Empire State Building. Without a chemical protectant, it is difficult to preserve cells during freezing. "If you flash-freeze a mammoth, that will basically fracture the entire cell, the DNA included," says Alex Greenwood of the American Museum of Natural History, an expert on ancient DNA. Adds his colleague Ross MacPhee, "I accept that a mammoth might be cloned if the material is good enough—that's a no-brainer—but the material is never good enough."

One prominent expert who thinks mammoth sperm might be salvageable is Ryuzo Yanagimachi, whose group at the University of Hawaii was the first, in 1998, to clone mice from adult somatic cells. That year, his group pulled off a feat perhaps more applicable to mammoths: they freeze-dried mouse sperm, reconstituted the powder, then inseminated female mice that gave birth to live offspring.

There may very well be freeze-dried mammoth sperm somewhere in Siberia. During the Great Ice Age, a dry, cold plain—the mammoth steppe—covered northern Siberia. It's not inconceivable that a juvenile male mammoth that died in midwinter, when temperatures may have plunged as low as −100 degrees Celsius, could have frozen to a depth of several inches within minutes. Such conditions would freeze his testicles before the semen or surrounding tissues had a chance to decompose. If the carcass lay exposed and untouched by scavengers, it would gradually dry out, a process that would happen quickly on a windy steppe. The mammoth would have become mummified, as do seals that die in the dry valleys of Antarctica.

Yanagimachi was one of a select group of researchers anointed by Buigues to try to clone a mammoth if suitable DNA—or freeze-dried mammoth sperm—were recovered.

The mammoth may not be the first extinct beast brought back to life. Efforts are now under way around the world to reanimate a menagerie of long-lost creatures. Probably the first one to be resurrected is a subspecies of goat called the bucardo. The last of these mountain goats, which once roamed the Pyrenees Mountains, died at Ordesa National Park in Spain in a freak accident (it was killed by a falling tree) in January 2000. Spanish researchers had sampled some of the female bucardo's tissues the previous year, and have since kept the cells alive in a nourishing broth. They too are working with Advanced Cell Technology to mount a cloning effort, and their odds of success are good, considering that the bucardo's cells are still alive and that many closely related kinds of goats are available to serve as surrogate mothers.

Several of the cloning candidates are from Australia and New Zealand, which together lost more than two dozen animal species during the colonization of those island nations over the last thousand years. A recent victim of this spate of extinctions could be the best candidate for cloning: the thylacine, or Tasmanian tiger, a marsupial wolf that resembled a dog with stripes on its back and an extralong, stiff tail. Despite supposed sightings of the shy animal in Victoria and other Australian states and territories in recent years, the last known member of the species died in a zoo in Hobart, Tasmania, in 1936, just weeks after the Tasmanian government finally granted the species protected status.

The Australian Museum in Sydney has kept a dead baby thylacine in formaldehyde since 1866. Now there may be a use for it. In April 2000, a group led by Don Colgan, chief of the museum's evolutionary biology division, announced that it had extracted well-preserved strands of DNA from the pup's heart, liver, muscle, and bone marrow. But the DNA must still be assembled into a complete sequence and then arranged into chromosomes, a feat beyond today's geneticists. Colgan has predicted it will take at least a decade to develop the technology to clone the thylacine, if it is indeed possible. They also must determine whether, as some suspect, the distantly related Tasmanian devil is the best living marsupial species to provide surrogate mothers.

A novel technique that may hold promise for the mammoth is being attempted in New Zealand on an extinct bird.

The huia bird was once prized for its white-tipped black tail-feathers. They were particularly fashionable in Europe as adornments for hats at the turn of the twentieth century—so much so that by the 1920s, no more huias were left to be shot and plucked.

Inspired by Dolly and the movie *Jurassic Park,* students at Hastings Boys High School in Hastings, New Zealand, wondered if scientists could bring back the huia, the school's emblem. They had the raw materials: a pair of stuffed huias on display in the school's reception hall. To discuss their proposal, the students organized a conference in July 1999 at which scientists, ethicists, and Maori, the indigenous people of New Zealand, discussed the ethics of such a project. A representative of the Ngati Huia, a Maori group that tried to save the bird in the early 1900s, argued that since humans drove the bird to extinction, they were obliged to try to resurrect it through cloning. That sentiment prevailed, and a team headed by Diana Hill of the University of Otago in New Zealand was chosen to lead the effort.

Hill acknowledges that the project faces daunting hurdles. The DNA will have to come from bone fragments or from stuffed huias, such as those at Hastings, or from better specimens in New Zealand's museums. That means the DNA is surely broken into pieces, and the huia genome, comprising about 60,000 genes, will have to be laboriously stitched together with the genes in the right places. If not, feathers may form on the feet or wings may grow out of the head, says Hill, who notes that she is pursuing this project as a hobby outside her regular research. Cloning a huia may sound like an impossible task, she says, but with the dramatic rate of progress in advanced gene technologies, anything is possible. The task would be about as challenging as piecing together a Phoenician urn that has broken into millions of pieces, some of which are missing. (Some experts believe that this approach is doomed to fail, pointing out that a single nucleotide in the wrong place can result in a lethal disease.)

The most promising strategy for the huia, Hill says, might be to catalogue the genes of a living relative, like the turkey or chicken, to get an idea of where comparable genes lie in the genome. If the researchers manage to put the huia's genome back together, they would insert the genes into artificial chromosomes, then inject the packaged DNA into the egg of a surrogate. That might be a magpie or a blackbird, or perhaps native New Zealand birds such as the tui or the kokako. She also hopes someday to bring back the moas, flightless birds that appear to have been hunted into extinction about 600 years ago. Her team has isolated growth genes from a moa thighbone preserved in a New Zealand swamp since the Dark Ages, but much of the rest of the DNA is shattered. "The most important outcome of this work has been to open vigorous debate within schools, the community, and the indigenous people on the use of advanced technologies by asking 'Can we?' and 'Should we?'" Hill says.

Another long shot—but one that might feed into any effort to revive the mammoth—is to clone a rare white elephant. These Asian elephants, with lightly colored skin around their eyes and trunks, are thought to bring good luck. Tissue from a white elephant owned by Thailand's King Rama III, who ruled from 1824 to 1851, was preserved in alcohol. Scientists at Mahidol University hope that cloning technologies will be sophisticated enough within a decade to extract the DNA from the precious remains and clone King Rama's white elephant using similar techniques to those for resurrecting a mammoth. (Some Thais believe that long-haired elephants—perhaps descendants of woolly mammoths—live today in the forests of northern Thailand. In December 2000 a group of conservationists led by Princess Rangsrinopadorn Yukol of the royal family embarked on a search for these fantastic creatures, which most observers believe to be nothing more exotic than Asian elephants.)

The tricky reproductive system of elephants could pose a formidable challenge to those who wish to clone mammoths. Researchers at the Smithsonian Institution's Conservation and Research Center have discovered that perhaps because of stress, many female elephants in zoos become "flatliners"—they stop ovulating. That's bad news for zoo populations of elephants, which some researchers predict will be reproductively dead in 50 years. Any female mammoth brought to life and kept in captivity may have similar problems ovulating.

But there are ways to fool the elephant's fluky reproductive system. John Critser has developed an ingenious "Horton Hatches the Egg" approach for growing elephant eggs. In the book by Dr. Seuss, Horton, an elephant, was tricked into hatching a bird's egg. Critser has tricked mice into making elephant eggs. In 1998 he transplanted frozen ovarian tissue from an African elephant into the ovaries of a female mouse; the elephant ovaries worked just fine, producing normal elephant eggs. This approach may be an alternative to the cumbersome job of using a long fiberoptic probe to collect eggs from an elephant—a task complicated by the difficulty in knowing when an elephant is ovulating. Critser's method can be used to produce lots of elephant eggs for in vitro fertilization experiments for the captive breeding of elephants—and perhaps someday for breeding mammoths.

This finding is particularly important because elephant eggs—without fail—are irreparably damaged when frozen in liquid nitrogen. But their ovarian tissue, Critser has found, is much more resilient. Since sperm freezes well, keeping ovarian tissue on ice will allow scientists to store both sex cells. Critser hopes to exploit this approach to produce elephant eggs, either for fusing with woolly mammoth sperm to breed a hybrid or for serving as a receptacle for mammoth DNA to grow an embryo of a mammoth clone.

Labs and zoos around the world are stockpiling sperm and tissue samples of rare animals, assembling a DNA version of Noah's Ark. Faced with the relentless degradation of habitats, scientists have little choice but to prepare these sperm and tissue banks for desperate attempts to bring species back to life. Saving as much genetic material as possible—even if that means keeping it frozen—is essential to the long-term health of a species. As a population dwindles in its deteriorating habitat, so does its gene pool. "The problem is that gene pools are being converted into gene puddles as the populations of species are reduced and fragmented. Gene puddles are vulnerable to evaporation," notes Thomas Foose.

Gregory Benford, a physicist by training who has won acclaim for his science fiction, such as *Timescape* and *Beyond the Fall of Night* (written with Arthur C. Clarke), has advocated an even more ambitious effort: to foray beyond the zoos and into the wild to salvage what he calls the library of life. In an article in the November 1992 *Proceedings of the National Academy of Sciences,* he argues the case for sampling species from disappearing habitats, such as tropical rain forests. The effort would freeze samples from enough individuals to provide the genetic diversity necessary to resurrect a species that will later go extinct. He described the rationale for such a project:

> Our situation resembles a browser in the ancient library at Alexandria who suddenly notes that the trove he had begun inspecting has caught fire. Already a wing has burned, and the mobs outside seem certain to block any fire-fighting crews. What to do? There is no time to patrol the aisles, discerningly plucking forth a treatise of Aristotle or deciding whether to leave behind Alexander the Great's laundry list. Instead, a better strategy is to run through the remaining library, tossing texts into a basket at random, sampling each section to give broad coverage.

Although scientists have begun mounting surveys of all species within defined areas, including one under way in the Great Smoky Mountains in Tennessee and North Carolina, Benford's frozen library of life is filled with mostly empty shelves.

Bringing back a long-lost creature is perhaps easiest if its genes persist in living animals. That appears to offer salvation for the quagga, a kind of zebra with a brownish hide and faint stripes on its hindquarters.

Quaggas were abundant on the South African plains until hunters wiped them out in the nineteenth century. The last known quagga, a mare, died in the Amsterdam Zoo in 1883. Over a century later Reinhold Rau, a taxidermist at the South African Museum in Cape Town, was remounting a poorly stuffed quagga foal when he discovered bits of dried flesh under the skin. He sent the tissues off to a DNA lab in California, which confirmed what he had long suspected: that the quagga was a subspecies of the plains zebra, an animal common to southern Africa. Working with game wardens, in 1987 Rau found several plains zebras with the most pronounced quagga-like features—brown tinged with faded stripes—and rounded them up into a breeding herd. The first members of the second generation of offspring were born recently, and they looked more like quaggas than their grandparents. The quagga gene variations are making a comeback by diluting the plains zebra gene variations in the animals' genomes. As exciting as it is to see the quagga's reclamation, it is equivalent to recovering a lost breed of dog—a task, in principle, that's far easier than recovering a lost species.

The new tools of reproductive biology give us the ability to fill the world with a menagerie of our design: from more-marbled cows to self-shearing sheep. Overriding natural selection and the survival of the fittest, people now control the evolution of farm animals. Will we do the same for wild animals, starting with endangered species? First we must decide which creatures share the land with us. That could relegate species, especially those considered charismatic, to a fate as sad as extinction. If pandas were to vanish from the wild, for example, pity the few survivors—perhaps clones of dead pandas—imprisoned in a bamboo reserve, sequestered from predators and under the constant watch of scientists who balance their diet and midwife their offspring. It might be a comfortable existence for the animals, but would our consciences accept it?

To some people, cloning an extinct creature is far more spiritually troubling than cloning a member of a species that's still alive. Some writers have conjured images of Armageddon. "The Holy Bible abounds with imagery of the dead rising from their graves when the end of the world approaches, and members of certain religious groups might view the resurrection of extinct creatures as the beginning of [the] Apocalypse," warned Corey A. Salsberg recently in the *Stanford Technology Law Review.*

Kazufumi Goto, however, argues that resurrecting extinct animals like the mammoth is morally acceptable. He says it would be easy to make the human race feel guilty, for hunters may have figured in the mammoth's demise. But that's not the force that drives him. Instead, he thinks that bringing back the mammoth will help people appreciate those species that have not yet disappeared from the earth and perhaps persuade them to put a higher value on sustaining these species.

In the summer of 2000, Goto was still waiting for his chance to test that idea. But he was no closer to launching a cloning experiment than he was when he first set out for Siberia in 1997. The Japanese team's most recent find was a disappointment. It took a full four months for Russian customs to allow their prize from the summer of 1999—a towel-size piece of skin from what they believed was a mammoth's rump—to leave the country. Later analyses by Iritani and his colleagues showed that the 30,000-year-old skin, as anticipated, contained only damaged cells with broken DNA—and suggested that it may have belonged to a woolly rhino, not a woolly mammoth.

Devastated, the Japanese finally began to accept what Sergei Zimov had warned from the beginning: you don't find frozen mammoths—they find you. So they aborted plans to prospect for mammoth tissue in 2000, instead putting up a bounty of a million yen (around $10,000) to anyone in northern Siberia who tipped them off to the discovery of well-preserved mammoth tissue suitable for cloning experiments. And, like everyone else who watched the Jarkov rise into the sky over the Taimyr Peninsula, they eagerly waited to see what would emerge from the permafrost block when it was defrosted.

PLEISTOCENE PARK

Imagining a baby mammoth brought into our world—so unlike the world its genes prepared it for—brings to mind a line from Rudyard Kipling's poem "The Story of Ung." In this Stone Age tale, Ung, a cave artist, "pictured the mountainous mammoth, hairy, abhorrent, alone." How lonely would a twenty-first-century mammoth be if scientists were unable to resurrect companions for it? And how pitiable if it were condemned to spend the rest of its life incarcerated in a zoo.

The problem is that its home is gone. Other than isolated grasslands on the cold, dry Tibetan plateau, perhaps, no place on the earth today looks much like the mammoth steppe, a vibrant ecosystem that dominated much of Siberia before vanishing after the Pleistocene Epoch ended around 11,000 years ago. This means that any new mammoth habitat would have to be created from scratch. Fortunately for scientists, just such an effort could soon begin. Sergei Zimov, the Russian ecologist who accompanied Goto and his colleagues on their trip to Duvannyi Yar, has devised a plan to recreate the mammoth steppe in a preserve in northeastern Siberia. His vision, Pleistocene Park, could be the best hope for a mammoth's survival in the wild.

Zimov showed me the beginnings of his park outside Cherskii on a cool August afternoon. Like a frog hopping from lily pad to lily pad, he strode from one tussock to the next, wobbling for a moment on each sedge knob rooted in the sodden permafrost. Occasionally he misjudged a tussock's firmness, and his leg disappeared up to the knee into the marsh water. Within minutes he reached higher ground and a carpet of mosses and lichens, birch bushes, willows, and scattered larches—hallmarks of a mixed tundra-taiga landscape that dominates much of northern Siberia above the Arctic Circle. It's a starkly beautiful, wild land, permeated with the fragrance of alpine sage. Zimov, however, wants to see it torn up and populated.

He pointed to a tangle of birch and willows a few hundred feet away, where two of the agents he hopes will carry out his grand scheme were picking their way across a ridge. They were young male Yakutian horses—off-white and pepper-flecked, the color of snow near a Moscow highway. Zimov envisions dozens of these horses, along with moose, reindeer, and a herd of bison imported from Canada, ripping up the mosses and shrubs with their hooves and teeth, allowing grasses to move in. Within a few years, he hopes, grazing animals will have supplanted the current ecosystem in a preserve roughly three times the size of Manhattan, with a grassland resembling one that existed here during the Pleistocene Epoch. "In some places we must not only preserve nature, we have to reconstruct it," Zimov says.

In creating Pleistocene Park, Zimov, two U.S. ecologists —Terry and Mimi Chapin, a husband-and-wife team at the

University of Alaska, Fairbanks—and a group of Canadian and Russian wildlife biologists are embarking on an ambitious experiment: to test theories about the forces that shaped, maintained, and ultimately vanquished a long-gone ecosystem.

Today's Pleistocene Park, however, looks nothing like a proper home for a mammoth. The land is dotted with hundreds of lakes, each less than half a mile across. The park's northern half, a floodplain with sediments laid down in the last several thousand years, is dominated by sedge meadows, dwarf shrub heath covered with lichens, and small patches of Arctic grasses that provide a glimpse of how the mammoth steppe once appeared. Muskrat paddle in the lakes while Arctic hares and voles burrow in the meadows. Predators, when they appear, include bears, wolves, lynx, wolverines, polar and red foxes, sable, and weasels. Every spring when the snow melts, the meadows flood for about two weeks, sending the park's denizens up the higher, forested ground to the south.

The southern section of the park is more ancient land: hills and terraces formed during the Pleistocene. While lakes in the northern floodplain are fed by precipitation and streams, this older land has mainly thermokarst lakes. They form when over many summers sunshine, little by little, melts ice wedges just beneath the surface of the permafrost. Eventually the soil on top of the wedges caves into the growing pool. A series of fires in the last thirty years has wiped out nearly three-quarters of the larch trees that once covered the park's southern half. Larch seedlings attempting to recapture lost ground vie with shrubs and willow bushes for nutrients.

Weather is at the crux of the debate over Pleistocene Park's success. Most experts argue that Siberia in the Pleistocene was much drier than it is today. They point to Pleistocene sediments, which contain pollen and other remnants of grasses that thrive in dry soil. These grasses are thought to have fed an array of large herbivores, including mammoths, steppe bison, horses, moose, reindeer, and woolly rhinos. "The ecosystem was just as vibrant as the savanna," Zimov says. The mammoth steppe once covered a huge swath of present Siberia, Canada, and Alaska, encircling the North Pole. After the Great Ice Age, this grassland gave way to tundra and taiga, while the proportion of permanently frozen muck—sand, silt, and loam—was halved to about one-seventh of earth's land surface.

Some scientists believe that a sudden and severe climate shift at the end of the Pleistocene—marked by a warmup of about 5 degrees Celsius in just 20 years and, perhaps, an increase in rain and snow—set in motion an inexorable turnover in which marsh-loving mosses and sedge conquered the circumpolar regions. But Zimov doesn't buy that. He and his staff at the Northeast Scientific Station have spent two decades probing Pleistocene sediments, and based on their observations, he argues that the rich Pleistocene soil appears drier than today's soil not because the mammoth steppe got less precipitation, but because grasses are much better than mosses at sucking water from the soil and releasing it into the air, a process called evapotranspiration. He points out that northeastern Siberia's climate is very dry now: his home base, Cherskii, receives around 7 inches of rain a year. Thus he believes that the Pleistocene soil simply soaked up more water than other scientists are willing to admit.

As a pilot trial to try to bring back grasses and banish mosses, in 1989 Zimov brought two dozen Yakutian horses about 240 miles east to Cherskii. This breed, the closest descendants of the horses that lived in Siberia during the Pleistocene, can put on a thick layer of body fat during the summer and fall that gets them through the harsh winter. Zimov kept the two-year-old foals for two weeks in a paddock, then allowed them to range within a 6-mile radius. Over three years he watched as the area in and around the paddock was gradually converted from tussocky shrubland to grassland. Then he set the horses free. To Zimov, the outcome was clear: if the horses could thrive untended and usher in grasses where they trod the taiga, then bison and other large herbivores might trigger an even more profound and lasting transformation of the land.

Terry Chapin recalls being intrigued but somewhat skeptical about Zimov's ideas when they first met in Oregon in 1991. But Zimov, who had developed a computer model that predicted how steppe or moss ecosystems might thrive under various ecological or climatic regimes, made Chapin a believer. In a paper in the *American Naturalist,* Zimov, the Chapins, and three colleagues argued that precipitation levels in the Pleistocene climate could have supported today's mosses just as well as they did the mammoth steppe.

What kept the steppe covered in grass instead of mosses, they say, were the big grazers. By churning up the ground with their hooves, bison and other heavyweights prevented mosses from gaining more than a weak toehold on the landscape. The grazers' dung provided fertilizer for the grasses that, in turn, nourished the animals. Aggressive hunting, they argue, decimated the big herbivores, and as the animals gradually disappeared, the ground was disturbed less and grew poorer in nutrients—conditions that could have ushered in mosses and accelerated the herbivores' decline. Zimov points out that mammoth steppe grasses persist today in areas of Siberia rich in nutrients—along rivers and streams, for instance—and have returned to areas where mosses were disturbed by buildings, roads, fires, and other human activities. "We don't have this ecosystem now for the simple reason that we don't have enough animals," he says.

Zimov hatched the idea of Pleistocene Park as a way to broaden his smaller experiment with the Yakutian horses. This time, however, the idea is to bring in a herd of bison to punish the mosses. Northern Siberia's climate is too harsh for North American plains bison or European bison, which need liquid water year-round. Instead, Zimov plans to import a plucky survivor from northern Canada, a rare subspecies called the wood bison (*Bison bison athabascae*). Able to suck down snow for its daily water intake, the wood bison should thrive in the park. The bison is also well adapted to the kind of bitter cold that grips Pleistocene Park, where temperatures average −27 degrees Fahrenheit in January. Despite the cold, the winter snow cover is usually less than 15 inches deep and is loose enough for animals to push aside to feed on the sedges and grasses beneath. The last snow comes at the end of May; then the region enjoys about three months of above-freezing temperatures before the first significant snowfall of autumn, usually in late September.

Larger, darker, and with a more pronounced hump than its southern cousins, the wood bison was presumed extinct until a small herd was discovered in 1959 in northern Canada, near the Nyarling River in Wood Buffalo National Park, which straddles the border of the Northwest Territories and Alberta. (Who knows what beasts may still be hiding in that desolate region—why not mammoths? "You'd think if only someone would look carefully enough, maybe they would find some still alive in the Canadian Arctic," jokes the overkill guru Paul Martin.)

The shaggy wood bison may be the closest thing we have today to the extinct steppe bison, the most abundant large mammal on the mammoth steppe during the Pleistocene. It's unclear when the steppe bison disappeared from Siberia. The most recent fossils found are about 5,000 years old, although a rock painting of a bison in southern Yakutia appears to be about 2,000 years old. Clearly they survived into historical times and were a favorite prey of human hunters.

Zimov and the Chapins have in mind a three-step plan for introducing wood bison into Siberia. First, about two dozen bison would be flown to Cherskii from a Canadian preserve, Elk Island National Park in Alberta. They would be acclimatized in a fenced section of Pleistocene Park. The Canadian government has agreed in principle to this plan so long as Zimov's team can raise the money to move the bison. Cherskii's weather should not perturb the bison; its conditions are similar to those northern Canada. And the forage is better: the vegetation that bison prefer to graze on—sedge meadows, willow savannas, and grasses—covers 70 percent of Pleistocene Park compared to less than 10 percent of the Elk Island National Park. The bison would be mostly young cows and a few young bulls to ensure "the maximum rate of reproduction," Zimov says. Within a few years, the growing herd would range freely.

Stocking Siberia with wood bison could benefit Canada, too. If disaster were to strike and the Canadian herd were wiped out by disease, the Siberian bison could be used to replenish the North American population. "If all the Canadian wood bison were to disappear," says Zimov, "we could give some back." Any bison shipped to Siberia would be screened for two diseases of primary concern, brucellosis and tuberculosis. It's unlikely that the bison would contract either disease in Pleistocene Park; the nearest cattle, eight cows in Cherskii, are free of the diseases, and there is no record of other potential carriers—sheep or goats—ever being brought to Cherskii. Herds of domesticated reindeer 50 miles north of Pleistocene Park are infected with a strain of brucellosis bacterium that does not appear to infect bison.

Several decades after the project began, the growing herd of bison would be allowed to roam the lowlands between Siberia's Kolyma and Indigirka rivers, a plain the size of France. Given that a bison can consume about 10,000 pounds of forage a year, the food available in the Kolyma-Indigirka basin—about 50 million tons per year—should support more than 1 million bison. The animals could easily steer clear of settlements: few places on earth, outside Antarctica, are emptier. Fewer than 24,000 people live in isolated villages throughout the region, more than half of them Yakuts and other indigenous people who scratch out a living breeding reindeer and horses.

Some experts call Pleistocene Park a watershed in efforts to study lost ecosystems. "It's a very exciting idea," says Martin. The experiment, he predicts, "is going to have a revolutionary effect on how we think about designing nature." The plan also rates high on aesthetic grounds. "It makes sense to reintroduce species that have been recently extirpated by human hunting or habitat encroachment," says Paul Koch, a specialist on Pleistocene mammals at the University of California, Santa Cruz. "It's just planetary hygiene."

Zimov's project is not the first to try to recreate the mammoth steppe. In the 1970s Savelii Tomirdiaro, a permafrost engineer, drained acres of lakes in Chukotka, a province east of Cherskii that sits across the Bering Strait from Alaska. He predicted that grasses would grow in the rich loam of the lakebeds. Indeed, meadows soon sprang up where the lakes had been, and horse and reindeer herders used the meadows for forage. But after several years the grasses gave way to a mossy bog. By making wildlife part of the experiment, "Zimov's daring project has much stronger scientific grounds than Tomirdiaro's project, which only pretended to rebuild the extinct tundra steppe grassland," says Andrei Sher, a Pleistocene expert at the Severtsov Institute of Ecology and Evolution in Moscow.

Although endorsing Pleistocene Park, experts point out shortcomings that could undermine it. The project's main goal—restoring the mammoth steppe—could be doomed because of missing elements: the namesake mammoths, of course, and certain climate features of the Pleistocene, such as cooler temperatures, less carbon dioxide in the air, and—perhaps—less precipitation. "Attempting to recreate the ecological scene from 10,000 years ago seems dicey," says Koch. "The climate was quite different than today." The bison could also succumb to an unexpected pathogen. Such a catastrophe occurred in the United States in the late 1980s, when the state of Maine tried to reintroduce woodland caribou into Baxter State Park. Within months all the caribou were dead, victims of a brain screw worm carried by white-tailed deer that jumped the species barrier.

The idea of Pleistocene Park may sound radical, but Koch and others believe it may not be radical enough. "If this is really going to work, they may have to try something much more daring and ecologically risky," says Koch. He predicts that bison and other grazers won't inflict sufficient damage to the mosses: "This region wasn't called the mammoth steppe for nothing."

Mammoths and woolly rhinos, Koch says, were more effective landscapers, clearing snow, uprooting vegetation, and knocking down bushes and small trees. It's possible that the Yakutian horses and bison will fail to disturb the environment enough to prevent mosses from recapturing the pastures, says Pavel Putshkov of the Institute of Zoology in Kiev,

Ukraine. But if so, it would provide strong evidence that mammoths were essential to the mammoth steppe—and that their fall may have been the first domino that triggered the loss of other herbivores in the Siberian Arctic. To maintain Pleistocene Park under this scenario, Putshkov says, "bulldozers could be used instead of mammoths to prevent pastures from becoming forests."

For Pleistocene Park to really work, Koch says, Zimov should introduce the closest living relatives of mammoths and woolly rhinos—Asian elephants and white rhinos, respectively. Koch acknowledges, however, that even if these species could adapt to the Siberian climate, "most contemporary ecologists and conservation biologists would become apoplectic at the thought of releasing exotic living organisms of this size and ecological consequence." While it may be too early to undertake such an effort, given the state of knowledge about these species, he says, "If we don't act soon we may not get the chance."

Martin certainly agrees. In 1999 he and his colleague David Burney of Fordham University wrote an article advocating the reintroduction of wild elephants into the American West. "Isn't this a heretical idea for those of us inclined toward deep reverence for the wild?" they asked. Replying to their own question, they pointed out that proboscideans—mammoths, mastodons, and early elephant relatives called gomphotheres—flourished in the Americas until they all died out by the end of the Pleistocene. Now, they say, Americans live "in a continent of ghosts." All that remains of the former presence of these animals are the fossils and the shrubs—mesquite, honey locusts, and monkey ear—that lured the proboscideans, which consumed the sweet fruits and dispersed the seeds. According to Larry Agenbroad, "Our ecosystem is not the same because there are certain plants that can only be propagated by going through a mammoth's gut." He believes that honey locusts, in particular, are dying out in the wild because they don't have mammoths to scatter their seeds.

People should embrace the notion of bringing back proboscideans, once a dominating force in the North American ecosystem, says Martin. "These are flagship species. They aren't daddy longlegs or fruitflies." Recent efforts to reintroduce wolves and bears in the southwestern United States are fine, he says, as these species indeed prowled there in historical times. But Martin argues for taking a deeper view of history and for considering how to restore the megafauna that once ranged across the continent. The giant ground sloths are gone, but perhaps rhinos could serve as substitutes. The same with mammoths: elephants would be their proxies.

"I'm a green patriot," says Martin, whose call to arms appeared in *Wild Earth* magazine. "I don't want to see my country sold short." He also wrote a revisionist take on the classic American song "Home on the Range": "I'd say give me a home where the mastodons roam, and the ground sloths and the glyptodonts play." (Glyptodonts were giant creatures like armadillos that went extinct at the end of the Great Ice Age.) The article paints a vivid image of elephants and bison dancing a "languid ecological minuet" on the American range someday. "I would introduce elephants symbolically in a reserve where they can function as proxies for the extinct animals," Martin says. Still, considering that most ranchers and politicians are against efforts to reintroduce the wolf in the West, it would require a revolutionary change of heart to build enough support for letting African or Asian elephants loose on the prairie. "People in this country have to stand up and say, 'Hey, we've got a continent here, a legacy that's more than we see in any of our natural parks.'" Martin says. "I think it will have to come from an unscientific, emotionally charged idea by the public at large."

Agenbroad thinks a good home for the mammoth would be California's Channel Islands. In a neat evolutionary twist, North America's enormous Columbian mammoths (*Mammuthus columbi*)—descendants of the southern mammoths that invaded the continent from Eurasia—gave rise to the smallest mammoths. More than 40,000 years ago, a group of Columbian mammoths appears to have swum the six miles or so then separating the southern California mainland and Santarosae Island. Perhaps they were enticed into the waters of the Santa Barbara Channel by the smell of plants or ripening fruits carried on a westerly breeze. Once on Santarosae, the animals remained, and over the course of generations they shrank—a common phenomenon on islands, echoing the evolution long ago of pygmy elephants less than a meter tall on Sicily and other islands in the Mediterranean.

Santarosae's Columbian pioneers gave rise to the pygmy mammoth (*Mammuthus exilis*), which stood less than 6 feet tall. (Some grown pygmies weren't much bigger than Shetland ponies.) The Santarosae mammoths are called pygmies, not dwarfs, because pygmies only have pygmy babies, whereas dwarfs can have normal offspring. The pygmy mammoth had a 30,000-year reign on the Channel Islands: the most recent bones date to about 11,000 years ago, a few centuries before the first evidence of human habitation.

Agenbroad dreams of restoring the mammoth to Santa Rosa. "That would be the place to put a small group of mammoths and see what happens," he says. They wouldn't shrink —it would take centuries for the island-bound mammoths to become mini-mammoths—but Agenbroad thinks they would thrive. "There's plenty to feed on out there," he says. Although the mammoth steppe is gone, plant species in the mammoth's diet still exist in isolated patches across the Northern Hemisphere. Keeping a cloned mammoth in clover would require extensive cultivation of these plants. If mammoths consumed food at a rate anything like modern elephants, an adult woolly mammoth (weighing, say, 13,000 pounds) would have to suck down 114,000 calories—about 500 pounds of forage—*every day!* They would gulp down food and water three out of every four hours of their life, producing a few hundred pounds of dung a day.

If the original diet were too difficult to duplicate, Agenbroad sees little challenge in designing a new formula. "We have people who can make cow chow and sheep chow and goat chow. They can make mammoth chow." But before launching such an experiment, he would have to seek permission from the U.S. National Park Service, which manages the islands. Getting approval would be a long shot, Agenbroad fears: "the Park Service would never stand for it."

Indeed, the mere thought of bringing Canadian bison to Siberia is raising eyebrows. Some experts worry that Pleistocene Park could be dangerously disruptive to the Arctic—particularly if the government of the Sakha Republic, which oversees Yakutia, allows the project to expand. If the bison and other herbivores successfully dig up the vegetation, the ice that makes up as much as 80 percent of the underlying permafrost will begin to melt. A preview of what might happen in Siberia is the damage inflicted in northern Alaska in the 1950s, when there were few restrictions on driving vehicles on the tundra. Even 50 years later, these tire tracks across the tundra resemble drainage ditches.

In the short run, disturbing the tundra's mossy carpet could produce a landscape so soggy that it might impede the establishment of grasses and threaten the survival of the bison and horses. Over decades, however, a Pleistocene Park across northern Siberia could spur global warming by unleashing millions of tons of carbon dioxide now sequestered in the permafrost.

Even without elephants and rhinos, Zimov's bold plan might seem a fantasy in the grim realities of Russia's downward financial spiral. Daniil Berman, an entomologist at the Institute for Biological Problems of the North in Magadan, Russia, says with affection, "Sergei is crazy to build Pleistocene Park in our economic situation." But he and others marvel at how Zimov has prevailed against the odds. Anticipating the inflation that would make rubles essentially worthless in the early 1990s, Zimov went on a spending spree, buying everything from wood for posts to a heavy tractor for clearing land for a fence. "Zimov has a brilliant speculative mind; on the other hand, he is a man of action," says Andrei Sher.

Zimov also won important allies in the Sakha government. Aware that Sakha could rely neither on subsidies from Moscow nor on revenue from its abundant but hard-to-extract nonrenewable resources—gold, diamonds, oil, and natural gas—its president Mikhail Nikolaev has embraced wildlife stewardship as a potential source of meat for the local population and, perhaps, tourist dollars. "In Yakutia, the density of animals is very small," says Zimov. "What will the Yakutian people eat?" In a few years the answer could be the wood bison and other herbivores that, he hopes, will live on his reconstructed steppe.

Although it may be difficult logistically for Russians—let alone foreigners—to travel to Cherskii, Zimov predicts that adventure tourists could boost the region's fortunes. He points out that every lion in a park in Africa brings in around $1 million a year in tourism. "I hope the density of animals in

Pleistocene Park in 20 years will be the same as in the Serengeti," he says. He envisions charter flights bringing in tourists from Alaska and Japan. Depending on how long it took to get through customs in Khabarovsk, the trip from either destination could be made in a day. The park, Zimov says, is strictly a nonprofit venture run by the Pleistocene Park Association, consisting of him and five other Russian ecologists. In addition to deeding the park's land to the association, the Sakha government has promised $1 million for the first 5 years after the bison arrive to cover expenses such as fences along the park's boundaries, feed to tie the bison over until the grasses take, veterinary services, and facilities for visiting scientists. And to allow researchers to monitor the animals in winter, Zimov plans to build a 30-meter-high observation tower on a barge that can be frozen in place on one of the rivers that wend through the park.

As a step toward making this fantasy come true, in May 1997 Nikolaev invited Canada to send wood bison to Siberia as part of a "major conservation effort for northern species and ecosystems." He emphasized that a Siberian herd "could serve as a breeding stock for future national or international bison conservation efforts." In a reply several months later, the Canadian environment minister Christine Stewart praised the plan—calling its scientific rationale "sound"—and gave permission to develop a plan for the bison transfer.

In spring 1998 Zimov took the first step, buying thirty-two Yakutian horses with funds from the Sakha government and taking them about 240 miles east to the park. Heeding the *Field of Dreams* mantra "If you build it, he will come," Zimov has fenced a section of marsh in which the bison will adjust to their new climate. His team is still looking for funds to cover the estimated $330,000 cost of readying, shipping, and acclimatizing roughly thirty bison, along with a veterinarian who would monitor the animals during their first few weeks in Pleistocene Park.

The scientists expect it will take several years for the animals to churn up and fertilize the ground enough for grasses—present in small patches throughout the park—to become widespread. Even now, Zimov says, the park has enough nutritious grasses, tasty willows, and sedge meadows to sustain several hundred horses, reindeer, and bison; in case of a hard winter or unforeseen circumstances, however, he has stockpiled enough forage to sustain 150 big grazers for a few years.

If the bison take to their new environment, Zimov has plans to reintroduce other animals that could make the ecosystem balanced and self-sustaining. To begin with, he plans to bring in musk oxen. In a Cold War gesture of goodwill, this Ice Age holdover—whose range had shrunk from much of the mammoth steppe during the Pleistocene Epoch down to isolated parts of northern Canada and Alaska in the twentieth century—was restored to Siberia's Taimyr Peninsula and Wrangel Island in 1974. Zimov would also like to bolster the ranks of predators. Already the park houses a wolf family, but he also hopes to add big cats, such as the Siberian or the Amur tiger, that would be surrogates for extinct cave lions and keep the wolves in check (otherwise the wolves would kill all the horses and reindeer, Zimov says). These tigers are threatened with extinction—their habitat shrinking and their prey scarce. The tigers would feast on the bison in Pleistocene Park, helping to cull the herd of sick or old individuals.

Rather than simply throw the animals together for an unfettered test of survival of the fittest, Zimov and his colleagues have devised a scientific plan that in the beginning would subdivide Pleistocene Park into various study units. Some areas would have each introduced species in isolation, others with various combinations. A few sections would be fenced off from all large animals. Inside one section, researchers themselves would tear up the ground to try to replicate the action of the grazers; inside another they would add fertilizers as a substitute for feces and urine; while a third section would remain untouched as a control. The fences would come down once the experiments were finished and the animal populations had outgrown their pens.

Besides supplementing the diets of the bison until the grass takes root, Zimov's team would likely have to seed areas, as grasses are scarce in the park. Creating a balanced ecosystem might also involve bringing in beetles, which make the fertilizers in feces more easily absorbed by plants as well as ground squirrels that spread grass seeds through their scat. Both animals exist throughout Yakutia, but only in small numbers. Apart from the bison and tigers—and perhaps mammoths, someday—all species brought to the park would be native to modern Yakutia.

Whether the tigers would be big enough to bring down young mammoths or elephants or rhinos (be they woolly or white) is another question. If not, says Koch, it would be up to people to hunt the hefty herbivores to keep the population in check.

Zimov realizes that he may never see mammoths cavorting in his version of the mammoth steppe. Indeed, he and others worry that a living mammoth might turn Pleistocene Park into a freak show. "There is the danger of having a park full of copies of the same individual," says Akira Iritani. To avert a dead end for the resurrected species, scientists would have to breed a minimum number of mammoths to support a population.

This recalls Noah's wisdom in taking aboard the ark seven

pairs of each species. Mammoth breeders may be able to use a skewed ratio: several females and a few breeding males. Each individual would have to have a unique set of genes—several clones of the same individual would lead to the persistence of bad or lethal genes, killing some offspring while deterring others from adapting to their environment. Such rogue genes are usually squelched or diluted in a healthy gene pool.

But before mammoths can gyre and gimbel in Zimov's Pleistocene Park, they must be summoned from the world of the dead. That was the hope of Iritani and Goto as they watched Buigues and his team begin the careful thawing of the Jarkov mammoth.

INTO THE ICE CAVE

For months I was forced to follow Bernard Buigues's exploits through secondhand reports from expedition members like Dick Mol and Larry Agenbroad. I had wanted to visit Khatanga myself, to see Buigues and his team in action. However, Discovery Channel—wanting the drama to play out in its documentary, not beforehand in the press—threw a shroud over the operations in Khatanga, prohibiting journalists not affiliated with the network from seeing the Jarkov airlift in person.

So I was forced to watch the documentary on television, along with 10 million other viewers. The program, which aired on March 12, 2000, was broadcast in 146 countries and was by all accounts a hit, outdrawing two immensely popular American shows, HBO's drama *The Sopranos,* and the wrestling program *WWF Smackdown!*

Discovery Channel could withhold the juiciest details about the expedition because Buigues, by contract, had ceded most of the control over publicity to the TV network. The other limiting factor was logistics. Because of Russia's restrictions on the flow of visitors to High Arctic towns, it would have been exceedingly difficult to fly to Khatanga on my own without written permission from the authorities there—and that, I knew, would have been difficult without Buigues's blessing.

Buigues and Mol had invited me to come in August 2000, when the expedition would fan out across the Taimyr Peninsula in search of new mammoth remains. They arranged for my official invitation from the regional authorities, which I used to get my Russian visa. I was about to buy my plane ticket to Moscow in mid-July when I got an e-mail from Mol saying that the plans had changed: Discovery Channel once again had vetoed the presence of any outside journalists. The producers preferred to control the dissemination of information about the expedition through the channel's Web site, hoping to save any big surprises for its next mammoth documentary, in March 2001. High-level talks between *Discover* magazine, which had commissioned me to write about the Khatanga expedition, and Discovery Channel failed to break the impasse.

It was becoming apparent that I would have to experience Khatanga vicariously through Mol and his colleagues. But in a surprising turnaround, in September the network decided to lift the veil for a few journalists—so long as we were escorted to Khatanga by a publicity manager for Discovery, Tim DeClaire. i was going to see the Jarkov mammoth at last.

On October 16 three of us—Gregory Feifer for the *Moscow Times,* Adam Goodheart for *Outside* magazine, and I for *Discover*—met DeClaire in Moscow. There we were joined by three of the expedition scientists, including Ross MacPhee, whom Mol had invited to join the expedition the previous spring—just the opportunity to return to Siberia that MacPhee had been hoping for. Buigues was already in Khatanga, waiting for us.

We were to fly to Khatanga via Norilsk, an inland Arctic city dominated by the mining company Norilsk Nickel. The city gained notoriety in the 1990s when environmental groups claimed that it was one of the most polluted places in the world.

With tickets in hand for the overnight flight, we were driven out to Moscow's Sheremetyevo One airport, only to learn that our flight had been suspended. Rumor had it that the financially strapped airline didn't have the funds that day to buy jet fuel in Norilsk for the return flight to Moscow. It was approaching midnight, and the other flight to Norilsk that evening, on another airline—KrasAir, based in the Krasnoyarsk region which includes Norilsk—didn't have enough seats available. That did not mean the plane was full; rather, airline officials realized that we were desperate to fly that night. Buigues's Moscow representative, a wily Georgian named Zurab Khvtisiashvili, did the negotiations by cell phone; once he rushed out to the airport a hundred-dollar tip per passenger for the airline officials, we got our seats.

Shortly after dawn we set down at Norilsk, with snow blowing across the runway. The stewardess announced that the temperature outside was −12 degrees Fahrenheit, then, repeating the information in English, lowered that to −13. The pilot then came on and said it was −14 degrees, which in his English monologue suddenly became −40. The safe conclusion was that after mild weather in Moscow, it was damn cold in Siberia. I shivered when our Ilyushin jet, after coming to a standstill, rocked from side to side in the strong gusts.

We had several hours before our flight to Khatanga. While

the scientists and a few Discovery Channel representatives were content to nap in the airport lounge, with its mustard-colored stucco walls and faded tile floor displaying the logo of the boycotted 1980 Olympics, we journalists did not want to miss what might be our only opportunity to see Norilsk —and make one last foray into civilization before heading into the wilds of the Taimyr Peninsula.

We hired a taxi for the forty-five-minute drive into the city. Paralleling the road for much of the way was the world's most northern railroad, a set of tracks deformed by the summer melting of the permafrost below a railbed used to haul smelted metal from the Norilsk mining operations to the airport. The electricity and telephone poles also had a hard time standing straight in the permafrost; many leaned cockeyed. The wind had subsided a bit, and as the sun climbed toward its low zenith—the days were rapidly getting shorter north of the Arctic Circle—ice crystals in the air, called diamond dust, cast a majestic halo. Two sundogs shot brilliant rainbow arcs, and a pillar of light stretched from the sun to the horizon.

The halo stayed with us for much of the drive, which passed through gentle, snow-covered tundra hills resembling a down comforter. The driver sped along the rough road, knowing precisely when to ease up on the accelerator to prevent the car with its balding tires from careening off the road. Of all his High Arctic adventures, Buigues later confessed to me that the drive from the airport into Norilsk was probably the riskiest leg of any trip.

Halfway there we passed through a small town, Nadezhda, meaning "hope." I wondered what the residents hoped for in this forlorn place, a collection of uninspiring concrete apartments clustered around a weathered wooden Russian Orthodox church. A tinier hamlet on the way had a chilling name in honor of the Stalin-era prison camp that it used to be: Kayerkan, "Valley of Death."

At last we entered Norilsk, passing through the city's gates, which bore an escutcheon with its emblem: a polar bear holding a giant key over its head like a weightlifter who had just completed a clean-and-jerk. We soon came upon a heavily guarded compound surrounded by a double iron gate around 25 feet high and studded with floodlights. Norilsk, we learned, was not only known for its nickel; it was also a major source of a far more valuable commodity, platinum.

On the outskirts, Norilsk looked like most any other big Russian city: block after block of drab Soviet-era buildings. But we began to perceive some differences as we were driven into the heart of the city. The buildings, many in pastel shades, all had a space between the bottom floor and the permafrost, a kind of aboveground cellar with holes in the walls for ventilation to keep the air trapped inside at the ambient temperature. That space prevented the buildings from melting the permafrost with their heat and subsiding. One building offered a double irony: a supermarket named Begamot, or "hippopotamus." A hippo was as unlikely an animal as you would ever find in the Arctic. Yet the name was an auspicious sign for our great adventure: the first western travelers in Siberia in the 1600s attributed the bones of mammoths to the biblical behemoth—the original meaning of *begamot*.

As we walked through the downtown, the wind kicked up again and penetrated our clothing, a harsh assault on bodies not yet acclimated to Siberian temperatures. At the end of the main street loomed the headquarters of Norilsk Nickel, an imposing edifice with black glass windows (maybe it was really the Borg ship from *Star Trek*) and a huge sign displaying the time. In block letters on a nearby building was the slogan *Slava Metalurgam!*—Glory to the Metallurgists! Looking down at our feet, however, we saw evidence of the price the residents paid for the glory of living in Norilsk: coal dust, including some thumbnail-size chunks, speckled the freshly fallen snow.

Having spent an afternoon in one of the most polluted cities on earth, we were ready for the tundra. As I had imagined, arriving in Khatanga after the three-hour flight northeast, our plane was greeted by the border patrol, which demanded to see our passports. One of Buigues's local fixers had come aboard the plane as well, to point out which passengers were part of his contingent and thus had a reason for being in Khatanga—and, most important, whose names were on a list of visitors approved by the local administration.

That evening we dined in Khatanga's restaurant—or Restaurant-Bar, as the sign on its corrugated metal exterior points out. Larry Agenbroad and the other scientists had steeled me to the uninspiring reindeer fare: ground reindeer topped by an egg sunny-side up. But to our surprise we were served far tastier dishes as accompaniments—Russian favorites such as marinated forest mushrooms and salted fish. As a bonus we got to hear a band tuning up for a wedding reception the next evening, crooning "Khatanga, Khatanga," a song with the kind of lounge-lizard melody that's unmistakable anywhere in the world.

Afterward, at the hotel Zapolyar'ye, "within the Arctic Circle," Buigues greeted us warmly. His eyes twinkled when he saw me, knowing how many false starts I'd endured before the Discovery Channel approved my visit. Before disappearing with the television producers, he addressed the three journalists in inimitable Buigues fashion: "You can ask Dick Mol about the science. You can ask me about women and love."

— — —

Over the next few days the TV producers—to the consternation of the journalists and DeClaire, who had to take heat from both sides—strove to keep us away from the filming, limiting our access to Buigues and the scientists. One explanation offered was that Buigues didn't act naturally with people watching. The latest contretemps gave us ample time to explore Khatanga, a town that may have more metal storage sheds than people. As in Norilsk, the snowy streets are flecked with coal dust from the electricity plant and several other industrial smokestacks in town. Few vehicles were on the streets. At one icy dip in a road, children were sledding on flattened cardboard boxes. Nearly everyone we encountered smiled warmly and offered an occasional hello—a bizarre phenomenon in Russia, where on the street one usually gets a scowl. Why are people so friendly? we wondered. It felt as if we were in the 1975 movie *The Stepford Wives,* in which women were replaced by cheerful and compliant robots. Goodheart, who in profile resembles Nicholas Cage, called Khatanga the Stepford Steppe.

We learned, however, that many of the Khatangans we met seemed happy because, in general, they are happy. Khatanga's economy is much more vibrant than that of many other towns in the Russian High Arctic. Buigues had done his share by spending heaps of money on leasing the town's helicopter, a fee that exceeded $1,500 per hour. But the town's major benefactor is Norilsk Nickel, which has operated mines and explored for minerals in the region since the 1970s, when the town was a base for diamond and oil prospectors. The company had helped pay for the schools and other infrastructure.

"Young people want to stay here; it's a good life," explained Sergei Pankevich, the deputy director of the Taimyr nature reserve, the largest Russian *zapovednik* in the Arctic. He took us for a tour one afternoon of the nature reserve's one-room museum on the first floor of an apartment building. Pankevich boasted that Khatanga had no organized crime, a scourge that had infected most of Russia.

At that moment a short Dolgan woman came over and disputed Pankevich's rosy take on life in Khatanga. "Life is very bad now," said the woman, whom Pankevich introduced as the museum's director, Yevdokiya Aksenova. She was wrapped in a black fur coat with a thick collar that seemed to support her chin; covering her head and her ears was a huge white Arctic fox *shapka*. "How to live?" she asked in a deep, throaty voice. "How to live?" Her stern look dissolved into a smile as she started telling us about a display of traditional Dolgan clothes: for winter, a thick cotton tunic and leggings decorated with beads of many colors and fringed with red fox. For summer, reindeer skin chaps and boots, also finished with beadwork. My attention was drawn to a wall with some clippings from local writers, including a stanza from the poem "Song of Khatanga" by Vladimir Kostyuchenko. It went like this (my apologies to the poet for the translation):

Khatanga of mine
White land.
Sunshine and frost,
The brilliance of the faraway stars,
Winds and blizzard,
Blue snows.
A light of hope
That illuminates the trail.

I was beginning to understand why people liked living here. I'm a snow freak, too, and love stargazing on a cold winter's night.

Aksenova pierced my reverie with a sudden laugh. In a

corner of the room she was showing off a few mammoth bones: a skull, a radius, an ulna, and a sensational neck vertebra still draped with cartilage and muscle. Since the 1700s, Khatanga has been a center of the mammoth ivory trade, she explained, hauling out roughly 50,000 tusks from the Taimyr Peninsula in the last few hundred years. "Nobody here cared about the bones," she said, "until Bernard came along."

These bones, it turned out, had a checkered history. After Dick Mol saw them in the museum for the first time in 1999, he asked Vladimir Eisner, one of Buigues's associates in Khatanga, about their provenance. In the summer of 1987, Eisner explained, a ranger found part of the head and shoulders of an intact mammoth protruding from the thawing muck on the Upper Taimyr River delta. The man hacked off the tusks and sold them—a legal enterprise—leaving the rest to rot. Fortunately, most of the carcass remained frozen underground.

Alexei Tikhonov later surmised that this same mammoth was at the center of a scandal in 1992. To celebrate the Year of the Mammoth in Japan, that autumn a film crew sponsored by Mitsubishi rediscovered the mammoth and hacked away at its neck with an ax, managing to carve off around twenty pounds of meat. "It was really barbaric," Tikhonov said. Further, the Japanese had not sought permission for the dig from Russia's Mammoth Committee, which is supposed to sanction all mammoth expeditions. After a flurry of angry letters, the two sides patched things up, and the Japanese agreed to sponsor a joint expedition the following summer to retrieve the rest of the mammoth. But water levels in the delta were high that year, forcing the team to abort its search for the submerged carcass.

At the end of August 2000, serendipity struck. Pankevich was scouting the delta for remains on behalf of Buigues, keeping an eye out for fresh signs of the Japanese team's mammoth. He wasn't having much luck, so sat on the banks of a stream to do some fishing. A few casts later his line snagged. Reeling in, he was amazed to find a tangled mat of golden brown mammoth hair caught on the fishhook. MacPhee came up with a name for the twice-rediscovered mammoth that stuck: Hook.

— — —

The next afternoon I got to see MacPhee and Mol in action in a one-room building housing a makeshift workshop and laboratory for the expedition.

Mol set a fragment of mammoth shinbone on a table. "Let's drill it," said MacPhee. The drill bit burrowed into bone, spewing smoke. Dousing the deepening hole with water to cool it hardly helped. "It's like it's on fire," MacPhee said as the room filled with the pungent odor of collagen, a sign of freshness.

The bones crowding the workshop's shelves included everything from bow-shaped mammoth ribs to the umber jawbones of extinct horses and even a wolf skull with small teeth resembling a dog's. They had been collected in August and September by eleven two-man Russian brigades that had scoured the Taimyr Peninsula for remains. Over the course of the summer, the team had hauled in more than a thousand specimens. Their trophies were quite secure: iron bars, reminders of the building's former life as a savings bank, fanned out like fish nets in the windowpanes. A gray cat, Masha, prowled among the bones, not the least bit interested in the scientists' activities.

MacPhee reached into one of the many pockets of his outdoorsman's vest and drew out a plastic sample tube. Using the back end of a paintbrush, he rammed the inch-long bone core out of the drill bit and into the container. From a total of fifty bones of mammoths, musk ox, bison, and horses, MacPhee took duplicate samples, one for dating the remains with high-precision atomic mass spectrometry, an expensive procedure that costs about $650 per sample, the other for DNA analysis to search for a hyperdisease pathogen.

His colleague Alex Greenwood will probe the bone samples for fragile strands of DNA that have not degraded. "Not knowing what we're after, it is totally like shooting in the dark," MacPhee says. Complicating matters, says Greenwood, is that "a lot of the really nasty candidate viruses that we might want to look at have RNA, not DNA." Going pathogen by pathogen could take years and end in disappointment, so they'll start their virus hunt by sweeping for suspects from broad classes such as the morbilli viruses, including the species-jumping canine distemper virus and Rinderpest, which at the turn of the twentieth century ravaged wildebeest, hartebeest, bongos, and other native African ungulates after jumping from Asian cattle. Another top suspect is an elephant herpes virus. "We know these viruses jump between African and Asian elephants and cause serious disease in elephants with very low levels of immunity," MacPhee says. "The best-case scenario," adds Greenwood, "is that we find a correlation, a specific virus showing up at the time of extinction that cannot be found prior to that time."

But the hunt won't be easy, as RNA is even more fragile than DNA. Since RNA degrades so quickly, Greenwood says, the best hope of finding an RNA pathogen in mammoth tissue hinges on finding viral coat proteins. Tough as the Energizer Bunny, these compounds form a sac around the virus's RNA. If the researchers do finger an RNA or even a DNA virus, they won't make any claims unless they score hits in samples from other recent mammoths as well.

MacPhee's group intends to collaborate with Jerold Lowenstein, an immunologist at the University of California, San Francisco, who was part of the group that isolated proteins from the baby mammoth Dima's blood. Greenwood came up with a plan to use antibodies to sniff out viral proteins lurking in mammoth tissue, and Lowenstein is just the person to pursue this strategy. MacPhee's group also might put promising bone slices under an electron microscope to hunt for viral particles lodged in the interstices. "With a hyperdisease pathogen, you should have jillions of particles in the body," he says. If the culprit were like rabies, which lays low in neural tissue, he admits that his team would never find it in bone. But the odds of finding a pathogen would rise if it were one of the viruses wrapped in a protective sheath called a capsid. "If it's something like flu or distemper," MacPhee says, "we stand a chance."

The Jarkov mammoth died too early to have been a victim of a hyperdisease bug, but MacPhee and Greenwood could still probe its tissues for signs of other microbes to eliminate them as hyperdisease suspects.

MacPhee had already retrieved some bone samples from Jarkov, which had been placed in an ice cave on the outskirts of Khatanga, but he and the other scientists were anxious to start thawing the block to see if there was any tissue inside. So were the journalists.

As the long polar twilight of early fall faded, DeClaire led us through an unfamiliar section of Khatanga and down a steep icy road. In the deepening dusk I could barely make out an ice fisherman's shack on a broad bend of the river. We hadn't quite reached the snow-covered beach littered with oil drums when DeClaire paused near what looked like an entrance to a mineshaft, a wrought-iron doorway in the hillside. He disappeared inside, then returned a moment later, nodding. As we followed him, I felt as if we were about to be initiated into a fraternity.

After my eyes adjusted to the dim lighting, I found myself in a house of horrors: stacked on pallets were dozens of dead bodies, beheaded and skinned. Here was the supply of the gamy meat we were served in town every day: slaughtered reindeers, a Clausian nightmare. We entered another tunnel and skirted a mound of stiff silvery fish. We were walking through the town larder. I could see my breath; inside the cave, temperatures hover around 5 degrees Fahrenheit year-round.

We entered an antechamber. Lined up against the wall was a row of exactly one hundred tusks of various sizes and colors —from pale white to reddish brown to bluish, a patina laid down by the mineral vivianite—that had been collected the previous summer. They awaited the practiced eye of Daniel Fisher, a tusk expert, who would drill out plugs of the tusks to ascertain the animals' health and the season in which they died.

But my eyes were drawn to a presence as commanding as the monolith in *2001: A Space Odyssey:* the gigantic block containing the remains of the Jarkov mammoth. Curling out from it were the mammoth's 9-foot-long tusks, hacked off the skull three years earlier by the Jarkov brothers. The majestic russet-colored tusks were anchored to the block by steel clamps—either out of respect for the mammoth or for a better television image, depending on whom you asked. The cave's arching, ribbed ceiling, covered with opaque ice crystals, made it feel as though we were inside the belly of a whale.

After it was airlifted to Khatanga in October 1999, the block had sat on the airport grounds, covered with a tarp, where it stayed frozen all winter and off-limits to the general public. "It was the best-guarded mammoth carcass in the world," says Mol. Then, in April 2000, Buigues came to an agreement with the owner of the ice cave and, using a crane and local manpower, dragged the 23-ton block on skids a half kilometer across town to its new home. "It was like a parade going from the airstrip to the ice cave," Mol says.

Now Jarkov was poised to become the best-studied mammoth in the world. In fact, this chilly meat locker would become a frozen laboratory for Jarkov and any mammoths hauled from the tundra in the future. "We'll have room for scientists from all over the world," Mol says. A nonprofit organization called Mammuthus, founded by Buigues, has been formed to oversee the studies.

Releasing Jarkov from his icy tomb would necessitate thawing the block millimeter by millimeter, a laborious process that could take months. But the scientific payoff of such a deliberate approach could be huge. Besides Goto's team, John Critser and Ryuzo Yanagimachi were hoping to lay their hands on frozen mammoth cells for a cloning attempt. Despite Mol's objection to cloning, he has remained generous. "We are not against scientists with a different opinion," he says. "Our policy is to give samples to everybody, not to say 'Hah! This is ours' and sit on it. If there's someone who wants to clone a mammoth, he has to wait until the moment we find good stuff," perhaps from the center of the block, which presumably has remained frozen for 20,000 years.

The time came at last to thaw the top layer of the block and unveil to the world the first remains of Jarkov. Emmanuel Mairesse, the director of the documentary, sat on the floor of the cave with his legs straight out, watching his crew adjust the lighting. Some light stands were propped up on white woven plastic bags of fish. Mairesse, a friendly guy with a dry sense of humor, had spent about sixty hours in the cave during stints in August and October. "You couldn't ask for a more perfect set," he said.

But he could have asked for a more perfect plot. Earlier in the day, Buigues had used a hair dryer to thaw the top few inches near the edge of the block. He found the tops of three spinal vertebrae, two of which were articulated—in anatomical position. But to his dismay, he also found a pair of ribs lying like crossbones, devoid of meat. Concern was mounting that Tikhonov, a year earlier, had been right in warning that there wouldn't be much of Jarkov to find. That was bad news for the television producers, but it didn't stop them from engaging in high theater. As the five stars filed into the cave—Buigues, de Marliave, MacPhee, Mol, and Tikhonov—they were told to don matching gray vinyl jumpsuits.

Out of earshot of the producers, we gleefully heckled them. "They look like Power Rangers," Feifer remarked. "Or like the guys from the Pentium commercial," I said. MacPhee cast a wan look our way. "I fear so," he said.

The klieg lights went on, and the actors, standing on a wooden catwalk, trained their hair dryers on the top surface of the block. Buigues hammed it up. "Did everybody here buy tickets for the show?" he asked. It was hard to imagine him discomfited by spectators, as the producers had claimed.

The chunk of ancient riverbank from the Taimyr Peninsula eroded speck by speck under the firepower of five Wigo Taifun 1100s. After a few minutes, the scientists turned off the dryers abruptly and huddled around Tikhonov, who was examining something. This wasn't in the script.

"It looks like muscle," said MacPhee, who wore a black headband that made the tall, bearded scientist look even more imposing in his costume.

"This is the first soft tissue we've seen!" exclaimed Mol.

They held their find up to the light: it was a few inches long and looked like a scrawny strip of beef jerky. At first the scientists didn't know what to make of the scrap, not attached to any bones or skin. But their faces soon fell. Whatever part of the mammoth it came from, the bit of shriveled tissue was a further blow, after the clean ribs, diminishing their hope of finding intact organs. This was not the exquisitely preserved carcass that Mol and the others were hoping for.

But for MacPhee, the meager find was more than he had expected. "I was among the skeptics; I didn't believe there'd be a shred of soft tissue." However, he said, "Naysayers like me were wrong—there might not be a lot, but there's not nothing." MacPhee believes that Jarkov might yield a complete skeleton, although it's unclear whether science would derive much benefit from another mammoth skeleton. And how would people react to the news that the Jarkov mammoth was little more than bones? "Will the public still be interested after being burned?" MacPhee asked. It might be better, he argued, to salvage at least part of the Jarkov mystique by leaving the block intact. "It has an iconic value," he said.

According to Larry Agenbroad, whose teaching commitments had prevented his being present at the Big Thaw, Buigues's legacy need not be the Jarkov mammoth. "I don't know what's in the block, and I don't really care," he said. The project was spectacular, he claimed, for demonstrating the possibility of keeping remains frozen as they are excavated. "Whether or not this mammoth is the epitome of frozen mammoths is immaterial," he said. "We can now go out and get more." So agreed the mammoth expert Andrei Sher, who pointed out that finding mammoth remains on the tundra by ground-penetrating radar is a "powerful new tool that we could only dream about before."

After the camera stopped rolling, Mol conveyed a similar message. Climbing out of his jumpsuit, he said, "We're here for the science, not for the film."

Still, the disappointment was palpable that night. "I was expecting a lot and got a little," admitted Buigues, looking worn. "I had in my head something quite alive. I smelled it; it smelled like a wild animal." But his uncharacteristic melancholy soon wore off, and his eyes brightened. "Three years ago, when I first started excavating Jarkov, this all was like a dream. This piece of meat does not shatter my dream. We cannot imagine what we'll find deeper."

— — —

The next day Buigues and his group went out on the tundra to survey Hook. Forbidden to tag along, we journalists continued our explorations of Khatanga, striking out late in the day on a road that led toward the outskirts of town. The slick concrete sidewalk ended, and we walked on the side of the road in the snow, so dry and cold that it squeaked like Styrofoam under our feet. The only car we saw was a Lada with a big Y on top—a student driver—who came tearing around the corner and nearly plowed into us.

The buildings became more and more decrepit; it was soon clear that this abandoned area had been military grounds. In front of one of the few compounds lit in the waning daylight, behind a barbed-wire fence, stood a teenage boy in uniform toting a Kalashnikov rifle and smoking a cigarette. Farther on, past a crumpled mesh radar dish, a chipped mural on one derelict building read: "In the army there is nothing more important to a soldier than discipline." I thought that at any moment we would stray a meter too far and the authorities would swoop in and apprehend us, like the big ball that gobbled up Patrick McGoohan when he attempted to escape from his prison community in the 1960s British television series *The Prisoner.*

The next morning I went out with the helicopter to pick up Buigues's team and to see Hook's vertebrae sticking out of the frozen mud. Unfortunately, the ground-penetrating radar that de Marliave had set up didn't work: it was too cold for the batteries.

As the helicopter was being loaded for the flight back to Khatanga, Alexei Tikhonov came over to me with a mischievous grin. He reached inside a Ziploc bag and pulled out a brown fibrous slab. The day before, he had used an iron pick to chisel into the concrete-like ground and remove a slice of muscle from Hook's right flank. At the camp the previous evening, as they were celebrating Boris Lebedev's fiftieth birthday with a tundra banquet of fish stew, bread, and vodka, Tikhonov said he couldn't resist passing around the mammoth meat for a celebratory bite. He didn't have any takers, so he tried some himself. Even after a few shots of vodka, he said, "it was awful. It tasted like meat left too long in a freezer."

Tikhonov asked if I'd like to try some. "This is one of the best-preserved mammoths ever found in Siberia," he said. I searched his face for signs of Agenbroad's so-called mammothitis. His lips were chapped from being out in the field in frigid conditions for two days, but otherwise he looked fine. Still, I didn't take him up on the offer. I wanted him to save that meat for the cloners.

EPILOGUE

If I live to see a woolly mammoth, I would say,
"Praise the lord!" I think it would be glorious.

— PAUL MARTIN

We are drawn to vanished and mythical creatures. Who, after all, would turn down a chance to see a live Tyrannosaurus rex or a saber-toothed tiger? Who wouldn't want to read an article in the *New York Times* about Sasquatch surrendering to the authorities? Or about a dragon taking up residence in an abandoned Scottish castle? Without being conscious of it, we weigh in our minds the odds of any of those scenarios coming true. We sort the real creatures—the dinosaur and the Ice Age predator—from the mythical ones. We categorize the mythical beasts too, lending more credence to those that could plausibly be real, and have evidence, however slight, supporting their existence. The supposed photographs and footprints of an apelike human are slightly more believable than tales of winged serpents.

What allows us to suspend our disbelief is that vanished and mythical creatures have, on occasion, jumped from our imaginations into reality. For example, a 5-foot-long fish called the coelacanth was presumed to have gone extinct 70 million years ago. One of the mottled dark blue fish, distinctive for its four odd flipperlike fins, was caught in a deep-sea net off the coast of South Africa in 1938; subsequently, the fish were observed in the wild. Here was a living relic: proof that a species thought lost to the world could suddenly reappear. Then there's the Vu Quang ox. Unknown to science, the 3-foot tall creature had existed only as a myth until remains of the animal were found in 1992 in Vietnam's primeval Vu Quang Nature Reserve. The tiny ox was the first new species of large mammal discovered since a forest giraffe, the okapi, was captured in modern Zaire in 1906. A live Vu Quang ox calf was captured in 1994, proof that even large animals could evade detection in the remotest corners of our "global village."

A woolly mammoth is no Vu Quang ox hiding out in the forests of central Siberia or in northern Thailand, or wherever people hope it might have found refuge. The mammoth will not surprise us like the coelacanth did. Don't expect to hear about one straying into an Arctic village or being hauled home as a Siberian hunter's trophy. This great hairy beast made its last stand in the northernmost reaches of Russia, on a frigid island, 3,700 years ago. Then the mammoth went extinct, depriving Siberia of the most marvelous creature that ever trod that windswept land.

Yet the mammoth lives on as the soul of Siberia. Its bones struck fear into the hearts of indigenous Siberians for millennia after it disappeared. It was a siren to waves of explorers who ventured into Russia's forbidding interior on missions to find out more about the beast and perhaps bring back its carcass. Although the mammoth was the first species ever shown to have gone extinct, it clings tantalizingly close to life, its flesh entombed in the Siberian ground. Perhaps the frozen mammoths, like cryogenic millionaires, are waiting for the right person to come along and breathe life into them. Kazufumi Goto hopes to be that person. He has the right combination of attributes for success: a childlike fascination with mammoths and a proven talent for reproductive biology. Now all he needs is a well-preserved mammoth carcass to reveal itself to him.

For Bernard Buigues, the mammoth is a kindred spirit. Having spent more than a decade of his life in the High Arctic, Buigues knows the feel in summer of knobby tussocks beneath his feet and the fresh scent of snow-covered larch trees in winter. He knows that in Siberia, a knife's edge separates triumph and tragedy, life and death. The airlifting of the Jarkov block was a tremendous gamble that paid off until the first look inside brought disappointment. But the experience was transforming for Buigues. An Arctic tour guide who seized the chance to sell a story about his mammoth-hunting adventures, he became a true mammoth lover who led the largest-ever expedition to gather the remains of Ice Age animals.

Buigues and Goto both hoped to find exquisitely preserved mammoths. They have failed—so far. But their quixotic quests have enriched our lives by bringing us closer to a part of our common past than we have ever been before. Their journeys offer visions of doom—and of resurrection. Buigues's team may not have unveiled a complete mammoth, but it has opened millions of minds to the riddle of the mammoth's extinction and the possible explanations for it, including a nightmarish plague. That possibility, in turn, raises a troubling question: if a creature as abundant as the mammoth can be erased from the planet, then isn't it possible that humanity could be snuffed out this way as well?

More questions swirl around any attempt to resurrect the mammoth. Would a living mammoth mock evolution? Or would the mammoth, rising like an icy phoenix from the Siberian permafrost, help ease our collective conscience, eliminating the vestigial guilt arising from the harm that our early ancestors inflicted on this species?

For Goto, reviving the mammoth would fulfill a pledge he made to the children in his hometown. Early in 1998, he gave up his faculty position to head a kindergarten his family owns. It was a wrenching decision, made only two years after he had earned a full professorship. But Goto says he wants to shape kids' minds well before university. "Occasionally I talk about the excavation to the children one-on-one. I hope someday that some of the kids who graduate from the kindergarten will do this kind of research. That would make me very happy," he says.

"Some people say, 'The mammoth is dead under the ice. Why don't you leave it in peace?' But I promised the kids I would create a mammoth," Goto says. "I want to do my best to accomplish the children's dream." Kazutoshi Kobayashi hopes to build on this dream at Field Stone Farm, an estate an hour outside Miyazaki. He will build a mammoth institute there, he says, if one of their Siberian expeditions meets with success. While any living mammoths would be kept at Pleistocene Park, exhibitions at Field Stone Farm would teach children about the mammoth and its place in a former—and perhaps future—world.

Agenbroad, too, believes his work is as much about the

next generation of people as it is about the next generation of mammoths. He and his son hope to arrange for the donation of computers to Khatanga so the children there can communicate with their peers in the United States via e-mail. He foresees a stream of adventure tourists going to Khatanga to see the frozen mammoths and other Ice Age specimens excavated by Buigues. "I don't think Disney's going to put a Disneyland out there," he says. However, "it doesn't take much to draw a billion bird watchers to Bali. Maybe the same thing can be done with mammoth watchers, where they see the animal with body parts still attached. The ice cave in Khatanga lends itself to that kind of experience." And if tourists don't want to go to Khatanga, Buigues could go to them. He's hoping to bring the jewels he unearths from the Siberian tundra to the rest of the world in a flying museum, a jet packed with the remains of mammoths and other exotic beasts.

A cloned mammoth would be a much bigger draw, of course. "Can you imagine the attendance at a zoo that had a woolly mammoth and a woolly rhino? This is beyond sporting event attendance," says Agenbroad, who notes that if a clone were born, he and Wanda, his wife, want to be the godparents.

As he contemplates cloning a mammoth, Agenbroad, like Goto, has tried to seize the moral high ground. "I hear some people say, 'Who do you think you are? You're playing God. You're going to bring about the demise of the world. Those mammoths are dead for a reason. Why do you want to bring them back?'" He responds that after humans arrived in the American Southwest, grizzly bears and wolves were extirpated; those species have recently been reintroduced to the region. "What's the difference between that and reintroducing a mammoth that humans probably also exterminated?" Agenbroad asks. "If it ever happens that we get a mammoth clone," he says, "I have my media statement ready: Na na-na na-na na!"

Agenbroad is hopeful that the Hook mammoth—which Buigues excavated in the spring of 2001, while the ground was still frozen, and transported by truck back to Khatanga—has the right stuff for cloning, a possibility only if its bottom half has stayed frozen in miraculously pristine condition since its death thousands of years ago. But Hook was not the only highlight anticipated for the Buigues expedition in 2001. Nor is cloning high on the agenda for Dick Mol, who will continue to lead the scientific program. Mol took a one-year leave of absence from his job at Amsterdam airport in 2001 to study the specimens the team has already collected and to participate in the summer fieldwork. His stewardship of the program—and his avowed skepticism toward the possibility of cloning the mammoth—has impressed other expedition members. "Dick is a very, very smart guy. But because he never got a position in academia, he has to wear this big A for amateur on his forehead," says MacPhee. He has appreciated the openness with which Mol has conducted the program, including a plan to distribute mammoth remains to anybody who submits a serious research proposal. Having earned the respect of his academic colleagues, Mol intends to make the most of his tenure as scientific director. "This is a once-in-a-lifetime opportunity," he says.

In the summer of 2001, Mol and the rest of the Buigues team planned to explore virgin territory for the Mammuthus project, the New Siberian Islands. This Arctic archipelago between the Laptev and East Siberian seas may be the best place in the world to find frozen soft tissue, says project scientist

Alexei Tikhonov, who notes that three mammoth legs, flesh and all, were recovered during Russian expeditions there in the mid-1990s. "It's a real mammoth freezer," he says. And, if MacPhee's hyperdisease hypothesis is right, these islands may be a new hot spot in the search for the incriminating genetic material of an Ice Age outlaw.

Frozen woolly mammoths are our strongest link to a time deep in our common past, before our own kind began writing, cultivating crops, or forming nations. These long-lost animals exert a pull on our minds that can be understood only through the eyes of the scientists and explorers in pursuit of the beast. As I journeyed through Siberia and got to know them, I grew to share their infatuation with a creature that looms so large in our imaginations.

I believe that a habitat can be created for the woolly mammoth, and that if we can find enough DNA to resurrect several breeding pairs, we would not doom cloned mammoths to a sad existence. Whether it is five years, five decades, or five centuries from now, woolly mammoths will once again walk the earth.

SUGGESTED READINGS

This list, intended to help readers further explore themes and sidelights in the book, is only a sampling from a mountain of mammoth literature that has piled up since Chinese scholars first chronicled the underground rat. My apologies for skimpy offerings in the chapters based primarily on interviews with the scientists and explorers and my impressions of Siberia. I've included some additional mammoth information for aficionados on my Website, www.frozenmammoth.com.

RAISING THE DEAD

Mammoth primers

No one has described more eloquently the mammoth's importance to paleontology than Henry Fairfield Osborn, "The Romance of the Woolly Mammoth," *Natural History* (1930), 30:227–41. A well-written synthesis of what was known about the mammoth at the turn of the twentieth century is Frederic A. Lucas's "The Truth About the Mammoth," *McClure's Magazine* (1900), 2:353–

59. Gary Haynes's *Mammoths, Mastodonts, & Elephants: Biology, Behavior, and the Fossil Record* (Cambridge University Press, 1991) is filled with enlightening comparisons between mammoths and their kin.

For the best scientific summary of what a live mammoth probably looked like, see this article by two of Russia's premier mammoth experts: "Exterior of the Mammoth," by Nikolai K. Vereshchagin and Alexei N. Tikhonov, in *Cranium* (1999), 15:1–48.

Mammoths in America

To glimpse Thomas Jefferson's fascination with mammoths, see his article: "A Memoir on the Discovery of Certain Bones of a Quadruped of the Clawed Kind in the Western Parts of Virginia," *Transactions of the American Philosophical Society* (1799), 4:246–60. Also fascinating in that volume is George Turner's "Memoir on the Extraneous Fossils, Denominated Mammoth Bones: Principally Designed to Show, That they are the Remains of More Than One Species of Non-Descript Animal," pp. 510–18.

Mammoths and other extinct or fantastic beasts figure prominently in Native American mythology. Two highly recommended summaries are W. B. Scott's "American Elephant Myths," *Scribner's Magazine* (1887), 1:469–78, and W. D. Strong's "North American Indian Traditions Suggesting a Knowledge of the Mammoth," *American Anthropologist* (1934), 36(1):81–88.

TREASURE OF THE WOODEN HILLS

Baby Dima

Much of Dima's story was drawn from interviews with the Russian scientists who retrieved and studied the baby mammoth. Additional observations can be found in Nikolai Vereshchagin's "Tracing the Mammoths," *Science in the USSR* (1984), 6:80–128.

And for more details on what one of the U.S. teams that studied Dima found, see Ellen M. Prager et al., "Mammoth Albumin," *Science* (1980), 209:287–89.

Early accounts of the mammoth

Western sources offer conflicting accounts of the centuries-old Chinese records on the mammoth. One recommended article is W. F. Mayers, "The Mammoth in Chinese Records," *China Review,* (1877), 6:273–76. Memoirs of Western emissaries to China provide evocative descriptions of mammoths; for example, see Eberhard Ysbrant Ides, *Three Years Overland From Moscow to China* (London, 1706).

The Adams mammoth

A rare account of the discovery of what might have been the most intact mammoth ever found—had the wolves not gotten to it first—is William Gottlief Tilesius von Tilenau, "On the Mammoth, or Fossil Elephant, Found in the Ice, at the Mouth of the River Lena, in Siberia," *Quarterly Journal of Science, Literature, and the Arts* (1820), 8:95–108. Easier to track down—and packed with information on several other important mammoth finds—is I. P. Tolmachoff, "The Carcasses of the Mammoth and Rhinoceros Found in the Frozen Ground of Siberia," *Transactions of the American Philosophical Society* (1933), 23:2–74.

The Berezovka mammoth

A delightful narrative on this arduous excavation (with rich details on indigenous Siberian culture) is E. W. Pfizenmayer, *Siberian Man and Mammoth* (London: Blackie & Son Limited, 1939). For a more sober account from the expedition's leader, see Otto F. Herz, "Frozen Mammoth in Siberia," *Annual Report of the Smithsonian Institution For the Year Ending June 30, 1903* (1904), pp. 611–25.

The Berelekh cemetery

The definitive account of this mammoth boneyard is Nikolai

Vereshchagin, "The Mammoth 'Cemeteries' of North-East Siberia," *Polar Record* (1974), 17:3–12. Also see Vereshchagin's "Tracing the Mammoths" (cited under "Baby Dima").

FIRST DESIGN THE KOBE STEAK

Breeding or cloning a mammoth

For a look at some of the science underpinning the Japanese group's dreams of resurrecting a mammoth, see Kazufumi Goto et al., "Fertilisation of Bovine Oocytes by the Injection of Immobilised, Killed Spermatozoa," *Veterinary Record* (1990), 127:517–20, and Akira Iritani, "History and Efficiency of Microassisted Fertilization in Mammals," *Bailliere's Clinical Obstetrics and Gynaecology* (1994), 8:1–12. A review of the legal issues surrounding mammoth cloning is Corey A. Salsberg, "Resurrecting the Woolly Mammoth: Science, Law, Ethics, Politics, and Religion," *Stanford Technology Law Review* (2000), vol. 1.

Mammoth make-believe

Marshall Gardner's self-published book on the earth's inner sun, *A Journey to the Earth's Interior*—intended as a serious challenge to the geological status quo at the time—is great entertainment. Another lighthearted read is Diana Ben-Aaron's April Fool's story anticipating the Japanese effort, "Retrobreeding the Woolly Mammoth," *Technology Review* (April 1984), 85.

RIVER OF BONES

Kolyma history

A fact-filled pamphlet in Russian that can be purchased at the Lower Kolyma Museum of the History and Culture of Northern Peoples in Cherskii, if one happens to be up that way, is "350 Years of Expeditions by Russian Explorers and Polar Navigators on the Kolyma."

THE RAT BENEATH THE ICE

Indigenous Siberians and the mammoth

The surviving mammoth folklore recounted throughout the book was derived mainly from interviews with indigenous Siberians and Russians. For a historical account of the Yukagir, see Waldemar Jochelson, "Some Notes on the Traditions of the Natives of Northeastern Siberia About the Mammoth," *The American Naturalist* (1909), 43:48–50.

North Sea fossil hunting

A taste of what Dick Mol and his colleagues have found off the coast of the Netherlands is John de Vos et al., "Early Pleistocene Mammalian Remains From the Oosterschelde or Eastern Scheldt," *Mededelingen Nederlands Instituut voor Toegepaste Geowetenschappen TNO* (1998), 60:173–86.

A DEADLY CHILL

Taken with the flood

Kudos to William R. Corliss and his Sourcebook Project for having the vision to reprint Sir Henry H. Howorth's rich encyclopedia of mammoth lore, *The Mammoth and the Flood* (London: Sampson Low, Marston, Searle & Rivington, 1887). Howorth fans can trace his arguments back to "The Cause of the Mammoth's Extinction," *Geological Magazine* (1881), 8:403–410.

Land of the mammoth

For the essentials on the environment of the mammoth steppe ecology, an invaluable—if difficult for the nonspecialist—resource is Valentina V. Ukraintseva, *Vegetation Cover and Environment of the Mammoth Epoch in Siberia* (Rapid City, S. Dak.: Fenske Printing Inc., 1993). A far more accessible book that folds interesting observations of the geology and ecology of the High Arctic into a story about the excavation of an Ice Age bison is R. Dale Guthrie, *Frozen Fauna of the Mammoth Steppe: The Story of Blue Babe* (Chicago: University of Chicago Press, 1990).

Overchill

A groundbreaking attempt to correlate the mammoth's physiology with the idea of a climate-induced extinction is Henri Neuville, "On the Extinction of the Mammoth," *Annual Report of the Smithsonian Institution 1919* (1921), pp. 327–38. The explorer Bassett Digby weighed in with "The Mystery of the Mammoth," *The Nineteenth Century* (1923), 94:222–32, followed by a delightful book, *The Mammoth and Mammoth Hunting in North East Siberia* (London: H.F. & G. Witherby, 1926). For an eviscerating critique of the overchill hypothesis, see Pavel V. Putshkov, "Were the Mammoths Killed by the Warming?" in *Vestnik Zoologii*, Supplement 4 (1997): 3–81.

KILLER WAVE OF THE NEW WORLD

Pleistocene extinctions

Many of the top minds on quaternary extinctions share their thoughts in a volume edited by Ross MacPhee, *Extinctions in Near Time: Causes, Contexts, and Consequences* (New York: Kluwer Academic/Plenum Publishers, 1999). *Discovering Archaeology* features a series of articles in its October 1999 issue on the debate over the possible causes of the mammoth's extinction.

Overkill

Compulsory reading are two articles by Paul Martin laying out this hypothesis: "Prehistoric Overkill" in *Pleistocene Extinctions: The Search for a Cause* (1967), ed. P. S. Martin and H. E. Wright, Jr., pp. 75–120, and "The Discovery of America," *Science* (1973), 179: 969–74. Larry Agenbroad weighs in with "Clovis People: The Human Factor in the Pleistocene Megafauna Extinction Equation," in *Americans Before Columbus: Ice-Age Origins*, ed. Ronald C. Carlisle, *Ethnology Monographs* (1988), 12:63–74. A novel twist—how drought could have conspired with hunters to vanquish the North American mammoths—is presented in C.Vance Haynes Jr., "Geoarchaeological and Paleohydrological Evidence for a Clovis-Age Drought in North America and its Bearing on Extinction," *Quaternary Research* (1991), 35:438–50.

NASTIER THAN EBOLA

Mammoth DNA

The first success at sequencing DNA from the nucleus of a mammoth's cell is reported in Alex D. Greenwood et al., "Nuclear DNA Sequences from Late Pleistocene Megafauna," *Molecular Biology and Evolution* (1999), 16:1466–73. For an overview of the biochemical processes that damage DNA, see Tomas Lindahl, "Instability and Decay of the Primary Structure of DNA," *Nature* (1993), 362:709–15.

Hyperdisease

Ross D. E. MacPhee and Preston A. Marx lay out their case for a pathogen's wiping out the mammoths and dozens of other species at the end of the Pleistocene Epoch in "The 40,000-Year Plague: Humans, Hyperdisease, and First-Contact Extinctions," pp. 169–217, in S. M. Goodman and B. D. Patterson, *Natural Change and Human Impact in Madagascar* (Washington, D.C.: Smithsonian Institution Press, 1997). For a rogues' gallery of diseases threatening wildlife, see Peter Daszak et al., "Emerging Infectious Diseases of Wildlife—Threats to Biodiversity and Human Health," *Science* (2000), 287:443–49.

Mammoths in miniature

The original article that rocked the rarefied world of quaternary paleontology, by Sergei Vartanyan et al., is "Holocene Dwarf

Mammoths from Wrangel Island in the Siberian Arctic," *Nature* (1993), 362:337–40. For an overview of the only true pygmy mammoths, see Larry D. Agenbroad's *Pygmy (Dwarf) Mammoths of the Channel Islands of California* (The Mammoth Site of Hot Springs, S. Dak., 1998).

THE BIG LIFT

The Mammoth Site

A lively illustrated description of North America's premier mammoth site is in Dick Mol et al., *Mammoths* (The Mammoth Site of Hot Springs, S. Dak., 1993).

THE DNA MENAGERIE

Frozen zoos

Gregory Benford's timely call for action, "Saving the Library of Life," appeared in *Proceedings of the National Academy of Sciences* (1992), 89:11098–101. Fleshing out this concept are contributions from David Wildt et al. in "Genome Resource Banks," *BioScience* (1997), 47:689–98, and Oliver A. Ryder et al., "DNA Banks for Endangered Animal Species," *Science* (2000), 288:275–77.

Cloning science

Two landmark reports are the Ian Wilmut team's first report of Dolly the sheep, "Viable Offspring Derived from Fetal and Adult Mammalian Cells," *Nature* (1997), 385:810–13, and Teruhiko Wakayama et al.'s mouse clones described in "Full Term Development of Mice from Enucleated Oocytes Injected with Cumulus Cell Nuclei," *Nature* (1998), 394:369–74. Cloned cows are pretty popular too: See Y. Kato et al., "Eight Calves Cloned from Somatic Cells of a Single Adult," *Science* (1998), 282:2095–98. In this rapidly moving field, several reports have described the health disorders often suffered by clones. For example, see Jean-Paul Renard et al., "Lymphoid Hypoplasia and Somatic Cloning," *Lancet* (1999), 353: 1489–91.

Cloning endangered species

One of the earliest post-Dolly discussions of this possibility is Jon Cohen's "Can Cloning Help Save Beleaguered Species?" *Science* (1997), 276:1329–30. Robert P. Lanza and his coauthors offer an enthusiastic overview in "Cloning Noah's Ark," *Scientific American* (2000), 283(5):84–89.

PLEISTOCENE PARK

Where the mammoths and the glyptodonts roam

Sergei Zimov and his colleagues outline the science of Pleistocene Park in "Steppe-Tundra Transition: A Herbivore-Driven Biome Shift at the End of the Pleistocene," *The American Naturalist* (1995), 146:765–94. To imagine a similarly revolutionary experiment in North America, see Paul Martin and David Burney, "Bring Back the Elephants!" *Wild Earth*, Spring 1999, 57–64.

INTO THE ICE CAVE

Hook mammoth

The first scientific report on this find—other than Alexei Tikhonov's taste test—was scheduled to be presented by Dick Mol at the "World of Elephants" 1st International Congress in October 2001 in Rome.

INDEX